AS & A Level Maths

FOR DUMMIES®

A Wiley Brand

by Colin Beveridge

FOR DUMMIES®

A Wiley Brand

AS & A Level Maths For Dummies®

Published by: **John Wiley & Sons, Ltd., The Atrium, Southern Gate, Chichester,** www.wiley.com

This edition first published 2016

© 2016 by John Wiley & Sons, Ltd., Chichester, West Sussex

Registered Office

John Wiley & Sons, Ltd., The Atrium, Southern Gate, Chichester, West Sussex, PO19 8SQ, United Kingdom

For details of our global editorial offices, for customer services and for information about how to apply for permission to reuse the copyright material in this book, please see our website at www.wiley.com.

For general information on our other products and services, please contact our Customer Care Department within the U.S. at 877-762-2974, outside the U.S. at 317-572-3993, or fax 317-572-4002. For technical support, please visit www.wiley.com/techsupport.

Wiley publishes in a variety of print and electronic formats and by print-on-demand. Some material included with standard print versions of this book may not be included in e-books or in print-on-demand. If this book refers to media such as a CD or DVD that is not included in the version you purchased, you may download this material at http://booksupport.wiley.com. For more information about Wiley products, visit www.wiley.com.

A catalogue record for this book is available from the British Library.

ISBN 978-1-119-07846-3 (pbk); ISBN 978-1-119-07847-0 (ebk); ISBN 978-1-119-07848-7 (ebk)

Printed and Bound in Great Britain by TJ International, Padstow, Cornwall.

10 9 8 7 6 5 4 3 2 1

Contents at a Glance

Table of Contents

Part II: Arithmetic and Algebra 71

Chapter 5: With Great Power Comes... 73

Chapter 6: Playing with Polynomials 93

Introduction

● ●

So, you've chosen to study maths for A level? Fantastic. Welcome to the course!

It's a bit of a cliché that A level is a big step up from GCSE – but only because it's at least partly true. That said, if you know everything from your GCSE, you could probably pick up about a third of the marks in Core 1 without doing any more study.

In reality, though, not many people know *everything* from GCSE, and if your teacher assumes you do when you don't, well, then you have a problem. Worse still, it's a problem that's up to you to put right!

Luckily, you've done something smart and got hold of a book to help you. You're not going to find exercises in it or nice pictures of Carl Friedrich Gauss, but you will find plenty of advice and instructions on how to tackle various types of questions.

As for me, I've been tutoring A level maths since 2008, so I know what students tend to struggle with – and the kinds of explanations that tend to make complicated ideas make sense. Throughout the book, I try to break everything down into simple, repeatable steps that will work even if the question is a bit different from the example I've chosen.

About This Book

This book is for you if you're about to start studying for your AS- or A level in maths – or if you're already doing the course and need to improve your grade.

I take you through the topics that you can expect to come up in the exams, starting with the GCSE knowledge you're expected to carry through to A level and ending with differential equations – which sound scarier than they really are.

The book is in five parts:

- **Getting started:** This part covers the things that will make your life easier if you know them before you start studying. If you're already some way through your course, they're still useful to learn or revise.

- **Algebra:** Here, you develop your algebra techniques further, building on what you learnt at GCSE to solve more-involved problems. The biggest new ideas are, I think, infinite series and logarithms.

- **Geometry:** Similarly, the geometry you need for A level builds on what you've done before. In this part, you take trigonometry to new levels and do more interesting things with vectors than you ever imagined.

- **Calculus:** For most students, calculus is a new topic (although if you've done a Further Maths GCSE, Additional Maths or even the IGCSE, you may be familiar with some of the concepts). Calculus is loosely the study of slopes and areas, and it's probably the most important mathematical concept you get from A level maths if you're planning to study a science at university. (For some subjects, statistical reasoning may push it close – but that's not in this book, so it's obviously not *that* important!)

- **The Part of Tens:** Lastly, I offer some quick hints and tips to help you get started on evil problems and avoid some of the mistakes everybody else makes.

Because the topics covered in the book are complicated enough, I've tried to keep the writing conventions simple. Here are the ones you should know:

- *Italics* are used for emphasis or to highlight new words or phrases in each chapter. They also indicate variables.

- **Boldfaced text** indicates key words in bulleted lists or the key steps in action lists. Vectors also appear in bold.

- Internet and email addresses appear in `monofont` to help them stand out.

Foolish Assumptions

Making assumptions is always risky, but it's unavoidable in this kind of book. Knowing where I'm coming from should help put you at ease, though. Here's what I've assumed:

- You've done reasonably well at GCSE, or (if you're coming back to maths after a break) you remember enough of your secondary school maths that you'd get at least a B. Don't worry, though – in Part I of the book, I review the skills you may have missed.

✔ You're competent with the basics of algebra (solving linear equations), arithmetic (you know how to add, subtract, multiply, divide and take powers of numbers), and geometry (you know the names of shapes, and you won't be surprised by the equations attached to them).

✔ You've chosen to do Maths A level, you have some enthusiasm for the subject, and you want to do well in it.

Icons Used in This Book

Here are the icons I use to draw your attention to particularly noteworthy paragraphs:

Theories are fine, but anything marked with a Tip icon in this book tells you something practical to help you get to the right answer.

Paragraphs marked with the Remember icon contain the key takeaways from the book and the essence of each subject.

The Warning icon highlights mistakes that can cost you marks or your sanity – or both!

You can skip anything marked with the Technical Stuff icon without missing out on the main message, but you may find the information useful for a deeper understanding of the subject.

Beyond the Book

In addition to all the great content provided in this book, you can find even more of it online. Check out www.dummies.com/cheatsheet/asalevelmaths for a free Cheat Sheet that provides you with a quick reference to trigonometric tricks and calculus rules.

You can also find several bonus articles on topics such as the link between arithmetic series and straight lines, and why the derivatives of the trigonometric functions are what they are, at www.dummies.com/extras/asalevelmathsuk.

Where to Go from Here

Head to Chapter 1 for an overview of what you can expect in your A level course. Or if you want to get your study habits straight first, Chapter 2 ought to be your first port of call.

You can also use the table of contents and index to find the specific areas you want to work on. This book is meant as a reference guide, so keep it handy with your study gear and look things up whenever you're confused.

I wish you all the best with your studies! If there's something that isn't clear, get in touch – I'm always happy to try to make things make sense. The best way to reach me is on Twitter (I'm @icecolbeveridge) or through my website (www.flyingcoloursmaths.co.uk).

Part I
Getting Started

In this part . . .

- ✔ Set off towards mathematical mastery.
- ✔ Put yourself in a position to do well on the exam.
- ✔ Refresh your arithmetic and algebra.
- ✔ Get your shapes ship-shape.
- ✔ Sketch graphs with aplomb.

Chapter 1

Moving towards Mathematical Mastery

*I*t's a big step up from GCSE to A level – especially if you're coming in with a B or a marginal A. The pace is pretty frenetic, and there's a fair amount of A and A* material from GCSE that's assumed knowledge at A level. If you're not especially happy about algebraic fractions or sketching curves, for example, you're likely to have a bit of catching up to do.

Luckily, this book has a whole part devoted to catching up with the top end of GCSE, as well as the stuff you'll need to learn completely fresh. In this chapter, I note where the content overlaps with GCSE and introduce you to A level algebra, trigonometry and calculus.

Reviewing GCSE

The good news is that if you've got a solid understanding of everything in your GCSE, quite a lot of Core 1 and a fair amount of Core 2 will be old news to you. Possibly less good news is that if you've got gaps in your knowledge, you need to fill them in pretty sharpish.

The four key areas where there's an overlap between the two qualifications are algebra, graphs and powers (all in Core 1) and trigonometry (which comes up a lot in Core 2). There are other bits and pieces, too – your arithmetic needs to be pretty decent in Core 1, where you don't have a calculator, and parts of Core 2 are likely to test your knowledge of shapes other than triangles – but generally speaking, these are the big four. The first part of this book is all about making sure you're up to speed with them.

Setting up for study success

Forgive me if you think it's patronising to tell you how to study – after all, you must have done pretty well with exams to get this far. I go into studying because A level is a much tougher beast than GCSE – it's possible for a reasonably smart student to coast through GCSE and get a good grade without needing to work too hard; by contrast, it's unusual to see someone glide through A level. And presumably, if you were finding it straightforward, you wouldn't be buying books like this to help you through it.

In this kind of scenario, everything you can do to optimise your working environment, your note-taking and your revision translates to quicker understanding and more marks in the exam.

Also, if you're studying at sixth form, there's likely to be a bit more going on socially than at secondary school. The more quickly you can absorb your studies, the sooner you can get out to absorb the odd lemonade with your friends. That is what you're drinking, isn't it?

All about the algebra

You've probably been manipulating algebraic expressions for years by now, and some of it will be second nature. However, just as a checklist, here are some of the topics you need to have under your belt:

- ✔ Solving linear equations (such as $7x + 9 = 3x - 7$)
- ✔ Expanding quadratic brackets (for example, $(x + 3)(2x - 5)$)
- ✔ Factorising and solving quadratics (such as $x^2 - 5x - 36 = 0$)
- ✔ Solving linear and nonlinear simultaneous equations
- ✔ Simplifying algebraic fractions

I recap all of these in Chapter 3.

All this algebra isn't just for the sake of jumbling letters around and feeling super-smug when your answer matches the one in the mark scheme (although that can be a nice motivator). Algebraic competence underpins just about everything in A level. Even in places where you'd normally expect to use only numbers (for example, Pythagoras's theorem), you may be asked to work with named constants (such as k) instead of given numbers (such as 3).

Grabbing graphs by the horns

Somewhat related to algebra are graphs. You rarely need to draw an *accurate* graph at A level; it's far more common to be asked to *sketch* a graph.

That's good news: sketching is much quicker and more generously marked than plotting. However, you can no longer rely on painstakingly working out the coordinates and joining them up with a nice curve. Instead, you need to know the shapes of several kinds of graphs you've come across at GCSE: the straight line, the quadratic and the cubic graphs as well as the reciprocal and squared-reciprocal graphs.

You'll frequently be asked to work out where a graph crosses either of the coordinate axes (which is really an algebra question), and you'll be expected to be on top of curve transformations. I cover all these in Chapters 4 and 10.

Taming triangles and other shapes

Triangles, obviously, are the best shape of all, which is why you spend so much time on SOH CAH TOA, the sine and cosine rules, and finding areas at GCSE.

Oh, and Pythagoras's theorem. If there were a usefulness scale, Pythagoras's theorem would be *way* off it. I can't think of a more important equation at A level, and you can read about it in Chapter 4.

Those skills are extremely useful in A level maths. Pretty much every Core 2 paper I've ever seen has used a triangle somewhere, and triangles frequently crop up in other modules. (If you're doing Mechanics, having strong trigonometry skills is a massive help.)

It's not just triangles you need to know about, though. Be sure you know the areas and perimeters of basic two-dimensional shapes like rectangles, trapeziums and circles as well as the surface areas and volumes of three-dimensional shapes such as cuboids and prisms. I recap these shapes in Chapter 15.

Attacking Advanced Algebra

As you'd expect, the algebra you're expected to do at A level gets a bit more involved than what you did at GCSE. It comes down to learning some new techniques and linking some new notation to ideas you may already have a decent grasp of.

You start with powers and surds, a GCSE topic that sometimes gets glossed over. You'll need to be fairly solid on these, as they come up over and over again in A level. In later modules, you have a calculator that will happily tell you that the square root of 98 can be written as $7\sqrt{2}$, but in Core 1, you need to be able to work that out on paper.

You also deal with sequences and series (extending the work you've done in the past), solidify the ideas of factorising polynomials, and do some work on functions – one of the most important ideas in maths.

Picking over powers and surds

Working out combinations of powers is one of the most critical skills for A level maths. I wouldn't say it's more important than topics like algebra, but a student's skill here is a strong indicator of how easy a student is going to find A level. If you're a bit rusty on the power laws, you're going to have to sort that out in fairly short order.

You also need to be pretty hot on your surds, especially in Core 1, when you're one calculator short of a pencil case. Throughout your course, you'll need to work square roots out in *simplified surd* form or, more generally, in *exact* form – examiners want to see things like $\pi\sqrt{2}$ rather than 4.443.

A step up from powers and surds are *logarithms*, which are handy functions for turning equations with unknown powers into equations with unknown multipliers. For example, without logarithms, $3^x = 100$ is hard to solve (you know the answer is 4-and-a-bit but not necessarily what the bit is), but with logarithms, getting the answer is a simple bit of algebra: $x\log(3) = \log(100)$, where log(3) and log(100) are just numbers you can get from your calculator.

Lastly, under the 'powers' heading, you need to work with one of the most interesting numbers in all of maths, *e*. It's a constant (exactly $1 + \frac{1}{1} + \frac{1}{2\times1} + \frac{1}{3\times2\times1} + ...$, or roughly 2.718281828459045...) with the lovely property that if you work out the tangent line of $y = e^x$ at any point, you find its gradient is equal to e^x.

Sorting out sequences and series

A *sequence* is simply a list of mathematical objects (often numbers, some-times expressions). A *series* is what you get if you add them up.

You probably did some work on sequences in the past (finding the nth term, for instance, or deciding whether a term belonged to a sequence), and that will stand you in good stead. At A level, though, there's a lot more to it (who would have thought?).

As well as the arithmetic sequences you know and love, there are geometric sequences (where each term is a constant multiple of the one before). There are also explicitly and recursively defined functions, which *sound* like a hor-rible wild-card but in fact are quite nice because you're told precisely how they behave in the question.

And there are binomial expansions, which are a really neat way of expanding expressions like $(1+2x)^6$ without needing to multiply out huge numbers of brackets. It's a particularly useful technique when you get to Core 4 and have to expand monsters like $\dfrac{1}{\sqrt{4+3x}}$ and use the result to approximate $\dfrac{1}{\sqrt{4.03}}$.

Of course, the binomial expansion is one of many things you do much more often in exams than you ever will in the outside world. (We have machines for that.) However, the idea of approximating things using polynomial series is a powerful tool for doing serious maths if you take the subject beyond A level. Oh, and it's handy for doing mental arithmetic tricks that make you look like a god, too.

Finding factors

I keep coming back to a theme in this book: things in brackets are (usually) happy things. In most cases, if you can put something in brackets – a qua-dratic expression, or a cubic, or a fraction – you almost certainly should. If you have to solve for where an expression is 0, the factorising makes it very easy; if you need to sketch a curve, the bracketed form is much easier to work with than the expanded one.

All the work you did in learning to factorise quadratics over the last few years will serve you quite well with this – although, as you may expect, you take it a few steps further at A level.

In Core 2, you learn to identify factors of cubics (and higher-degree polyno-mials) using the factor theorem. You use polynomial division to take these factors out so you can factorise the remaining expression. You also use its cousin, the remainder theorem, to find out what's left over without having to work through the whole division.

Sometimes, though, you need to do the whole division or, depending on your preferred method, find a way to work around it. I like to turn the problem on its head by coming up with a template answer and seeing which numbers have to go in the template, but your mileage may vary.

All this factor fun shows up in Chapter 7, along with the Core 4 topic of partial fractions. Since you started working with fractions, you've been adding and subtracting them using common denominators. Partial fractions is the reverse process of taking a fraction that's been combined and splitting it up into the parts that once made it up. Why would you do such a thing? Two reasons: it makes things much easier to integrate, and it means you can apply the binomial expansion much more easily.

Functions

A mathematical *function* is, roughly speaking, a recipe for taking one or more values and spitting out another. They're a big deal, mathematically speaking: being able to talk about functions in the abstract, without explaining what the recipe is, means you can do interesting things with graphs and calculus without getting bogged down in the details. For example, if you compare the graph of $y = f(x-2)$ with the graph of $y = f(x)$, you can say, 'The graph has moved two units to the right' without caring whether the function is quadratic, reciprocal, trigonometric or other – quite a handy trick!

In Chapter 9, you learn about the slightly esoteric notation you use for defining functions. You find out how to combine functions with each other, how to *invert* (undo) functions and how to solve equations involving functions.

You also get to play with *iteration*, which falls into the dull-but-usually-straightforward category. The idea is that you set up a recursive process, doing the same thing over and over again, until it converges on a specific value.

Getting to Grips with Geometry

Geometry – literally 'measuring the Earth' – has developed over time to mean the study of shapes. At A level, the most important shapes are triangles (clearly the best shape) and circles (which are really triangles in disguise), although you will need to deal with rectangles and trapeziums and all manner of three-dimensional shapes in good time.

The four main areas of A level geometry are

- ✔ **Coordinate geometry,** which is about dealing – as you might expect – with coordinates; that includes midpoints, distances and equations of curves
- ✔ **Circles,** including their equations and some theorems
- ✔ **Triangles**, including advanced trigonometry
- ✔ **Vectors,** including vector lines, angles between vectors, and triangles in three dimensions

Conquering coordinate geometry

Coordinate geometry is a big topic at A level. You need to be super-confident with your *x*s and *y*s. You've covered the basics at GCSE – the equation of a line, finding midpoints and distances between points using Pythagoras's theorem, and so on – but it all gets taken a bit further at A level.

There are curves to sketch and shapes whose areas need to be known. After you do some differentiation, there are tangents and normals to find the equations of.

Setting up circles and triangles

I'm always surprised when I ask for the equation of a circle and someone says, 'πr^2'. First of all, that's not an equation (he or she means $A = \pi r^2$, where A is the area and r is the radius), and second, that's the *area* of the circle, not the circle itself.

The equation of a circle, like the equation of a line, gives you a relationship between its coordinates. If the equation is $(x-3)^2 + (y-4)^2 = 25$, one of A level's favourite circles, you can tell whether a particular point is on the circle by putting its coordinates into the equation as x and y.

In Chapter 11, I show you how to work out the equation of a circle as well as work out the area of a sector, the length of an arc and other things related to a circle.

Circles are closely linked to triangles; apart from all the trigonometry you know from GCSE, you'll also need to be able to do it all in *radians*, a much better measurement of angles than the degree. Fortunately, much of it is just a case of switching your calculator mode and relabelling your graphs.

Taking trigonometry further

If you split the word *trigonometry* up into its parts, you get 'tri', meaning three; 'gon', meaning 'knee' or corner; 'o', which means nothing; and 'metry', meaning 'measuring.' Trigonometry is about measuring things with three corners.

However, that's not all you use it for. It also has applications in any situation where things are periodic – measuring tides, modelling daylight lengths and analysing sounds, just off the top of my head. For that reason, you need to be able to take trigonometry further. Some of the things you'll be doing include

- Finding all the possible solutions to simple trigonometric equations in a given interval
- Using trigonometric identities to turn trigonometric equations into something you can solve
- Exploring what happens to sine, cosine and their friends when you add angles together
- Adding sine and cosine waves together to get another sine or cosine wave
- Working with the minor trigonometric functions – the reciprocals of sine, cosine and tangent, which are the cosecant, secant and cotangent (respectively); usually, they're denoted $\csc(x)$, $\sec(x)$ and $\cot(x)$

You also need to be up to speed on proving that two trigonometric expressions are equivalent. That generally involves combining fractions, applying identities, factorising things cleverly and understanding the symmetries of the various functions. Because these problems bring so many areas together, they're one of the most demanding (but also most rewarding) bits of Core 3.

Vanquishing vectors

You've done some work with vectors at GCSE – although you may not have done as much as you'd like, because vectors are usually in the A* questions at the end of the paper. It's OK, though: your GCSE vectors work isn't essential to your A level studies.

Vectors at A level can seem intimidating, with new vocabulary, new ways of multiplying things together, new equations of lines. After you get past that and think about how vectors fit together, they're really powerful – and straightforward.

In Chapter 13, I show you the following:

- ✔ How to come up with the vector equation of a line
- ✔ How to find the distance between two points (the length of a vector)
- ✔ How to find the angle between two vectors
- ✔ How to tell whether two vector lines cross
- ✔ How to find a point on a line so that a vector involving it makes a right angle with the line

Additionally, in case you're studying for the OCR MEI board, I take you through the gory details of vector equations of a plane – usually a big part of a Section B question in Core 4.

Conquering Calculus

In terms of A level maths, calculus is pretty much what you've been building towards for your entire career. The discipline has plenty of rules to learn but is based very carefully on much of what you've done up until now in algebra, arithmetic and geometry.

You can think of calculus as the study of curves, very loosely speaking: How steep are they? How much area do they have underneath them?

Why is that important? In many processes in physics, chemistry, biology, economics, psychology and anywhere else you care to apply maths, either there's a relationship between how much of a thing there is and how quickly it's changing, or you're actually interested in how something is changing. For example, your speed is how quickly your position is changing, and your acceleration is how quickly your speed is changing.

To work that out, you *differentiate* an expression or an equation. *Integration* is the reverse process. And because Core maths isn't just about doing maths for the sake of it, you need to know how to apply calculus to situations, both in mathematical and real-life contexts.

Dashing off differentiation

Differentiation is the process of finding how a quantity is changing instantaneously – at a specific point in space or time. It's kind of a big deal. Two of the greatest scientists of the early eighteenth century, Isaac Newton and Gottfried Leibniz, squabbled for years over priority (Newton thought of it before Leibniz, but Leibniz published first, leading to Newton having a hissy fit and accusing Leibniz of pinching his ideas).

The history of calculus is interesting but not especially useful for understanding what to do. In the chapters on differentiation (14 and 15, to be precise), you learn the following:

  - ✔ How to differentiate powers of x (or any other letter you put your mind to)
  - ✔ How to differentiate the trigonometric functions $\sin(x)$, $\cos(x)$ and $\tan(x)$
  - ✔ How to differentiate exponential and logarithmic expressions
  - ✔ What to do when expressions are multiplied together, divided by each other or applied to each other

Here, you also learn about the geometrical interpretation of differentiation – how to find tangents and normals, how to determine and classify turning points of a function and how to find whether a turning point is a maximum or a minimum.

Inspiring yourself to integrate

Integration is the reverse process of differentiation: taking a gradient and turning it into a curve. It's also used for finding the area underneath a curve. In fact, all the formulas for the area and volume of shapes can be worked out by integration, but I leave that as an exercise for the very interested reader.

In Chapters 16 and 17, I show you how to

  - ✔ Integrate powers of x (apart from that pesky x^{-1})
  - ✔ Remember to add a constant of integration
  - ✔ Work with limits to evaluate definite integrals and areas
  - ✔ Find the area between curves
  - ✔ Deal with trigonometric functions such as sine and cosine
  - ✔ Work with exponentials and (at last) that problematic $\frac{1}{x}$
  - ✔ Manage functions of functions and expressions multiplied together

Applying the calculus

There are two main schools of thought when it comes to maths. The crazy pure mathematicians think maths should be done for its own sake, for the sheer beauty of it, because it's fun. The more sensible applied mathematicians do maths because it's useful.

And calculus, while it's beautiful and fun, is especially useful. It's useful geometrically: it allows you to draw tangents and normals, find turning points and classify them, and that sort of thing. But it's also useful in what exam boards think of as real life, too. Here are some examples:

- **Finding maxima and minima of functions:** This allows you to find the most efficient way to use material to package products, or at what price to sell something, or how to minimise the cost of a journey.

- **Finding rates of change:** Knowing how quickly your experiment is running or at what speed your car is going is often useful.

- **Finding areas and volumes:** You may need to know how big an irregularly shaped object is. For a slightly more abstract application, the area under a speed–time graph gives you the distance something has travelled.

- **Solving differential equations:** In many applications, the value of a function is linked to its rate of change. For instance, the amount of a radioactive substance left depends on how quickly it decays, and the position of a satellite depends on its acceleration. Solving the links between derivatives and values is a key skill, both in Core 4 and as you take maths further.

Chapter 2

Setting Yourself Up for Study Success

· ·

In This Chapter

▶ Getting the right equipment and people

▶ Believing in your abilities

▶ Creating a study plan

▶ Using revision techniques

· ·

*P*reparation is nine tenths of the battle for A level maths (and, for that matter, most other things). Your goal between now and the exam should be to put yourself into the best possible position so that whatever they throw at you, you have the best possible chance of doing yourself justice.

In this chapter, I go through the details of *how* to study – what equipment you need, what you need from your workspace and what you need from anyone you choose to work with. I also talk about the mental-resilience side of studying (it's a hard slog, and for most people, studying maths isn't a smooth ride). You need a good attitude, good perseverance and a good way of talking to yourself – as well as good recovery when you mess up.

Messing up is *totally normal* and part of learning maths. If you're not making mistakes, you're probably not challenging yourself enough. I made mistakes in writing this book, but I (or, more often, my minions at Dummies Towers) caught them and put them right.

The final things you need are the self-reliance and motivation to come up with your own study plan. You *can* blindly accept that your teachers know how you learn best, but the chances are that they're working in an order and at a pace that they judge will be okay for most people – not necessarily optimal for you! If you take control of what you learn when, you can get significantly ahead of the class and lose some of the 'I'm not sure we've covered this' stress.

Equipping Yourself

I interpret *equipment* very broadly in this chapter. It's not just the mathematical paraphernalia like calculators and sharp pencils but also your whole studying environment: where do you work? Who do you work with?

You're probably tempted to say, 'I've got my calculator, I've got my desk, and I work *alone*, dammit' and skip straight to the next section – but I reckon it's worth making sure those things are well-suited to what you need. Sometimes, they can be the difference between an easy homework and a hard homework (or between a hard homework and an impossible one).

Stuff you need

In terms of actual equipment, there's not a whole lot you need for A level maths. A supply of pens and paper. Some pencils for drawing. A good scientific calculator (the calculator tips in this book are geared towards the Casio FX-83 or -85, but other brands are available).

You want somewhere to store your notes – ring-binders and poly-pockets are a good way to do that, although I advise against carrying ring-binders around with you, because the rings tend to break. They're a storage solution, not a transport solution! Alternatively, notebooks are always an option.

Set squares? Nope, definitely don't need them. Compasses? Nah, not really, unless you want your circle sketches to look really neat. Rulers? Again, the only time you *need* a ruler is if you're drawing something accurately, and there's very little of that at A level. If you want your axes to look neat, I suppose there's not much harm in rulers, but I personally make a stand for proud scruffiness in sketches.

Possible study aids to consider are

 ✔ **Index cards:** Index cards (bits of cardboard, typically about 8 centimetres by 13 centimetres) are really popular in the USA for revising but are bizarrely underused here. Write a question on one side, the answer on the other and presto! A quick quiz so you can revise whenever you have a spare moment.

- ✔ **Coloured pens:** Some students find it easier to remember their notes if they're colourful. If you're one of those, treating yourself to nice felt-tips is a good idea.

- ✔ **Textbooks:** I know, I know, you've just bought a brilliant *For Dummies* book! However, the official textbooks for your course have questions of the specific style you can expect in your exam as well as precise details of what you're expected to know.

Where to work

The best place to work is wherever you feel comfortable working. As a teenager, my best place to work was spread out on a bed (which at least forced me to tidy up before going to sleep). You might prefer working at the kitchen table, or at a coffee shop, or on the bus – or even changing it up and never working in the same place twice. Here are some things to consider about your work environment:

- ✔ **Lighting:** It's written into the parenting contract that, at some point in a child's life, a parent will have to turn a light on and nag, 'You'll hurt your eyes.' I don't believe you will hurt your eyes if you work in poor light, but you will make more mistakes. It's harder to read the questions and your work if the lights are too low.

- ✔ **Noise and distraction:** If you can work in a noisy environment, I envy you. Go ahead and work in the café or at the station. If you find it difficult to concentrate when there's hubbub, bear that in mind – close your door, ask people to leave you alone, and/or put on headphones to close out distractions.

- ✔ **Music:** Talking of headphones, try working with and without music. I find music with lyrics distracts me, but classical or instrumental music keeps me focused. You can also find noise generators that give you a background hiss to drown out noise.

Who to work with

Every so often, I think about education, exams and what the system teaches, and I cry a little bit: the whole system seems to be set up to show you that maths (and other subjects) are individual pursuits – working with other people is somehow cheating.

No it [expletive deleted] well isn't!

Maths is a collaborative effort. Pythagoras had a school of mathematicians he worked with. The boffins at Bletchley Park figured out the maths behind computers and codes as a team. Euler, probably the greatest genius in all of maths, collaborated widely. For every Andrew Wiles, who proved Fermat's Last Theorem more or less on his own, there are ten Paul Erdöses, who worked with everyone he could find, and a thousand unheard-of mathematicians who tried to do stuff on their own and failed.

The point is that doing maths on your own is much harder and much less fun than doing it with friends. There are all sorts of advantages:

✔ **Several heads are better than one.** Between you, you can figure out the answer to problems much more efficiently than on your own.

✔ **You're good at different things.** You can learn from each other and really shore up your own knowledge by helping each other.

✔ **You're building teamwork skills.** Employers like that. Especially in maths, you'll stand out.

It's important to pick the right people to work with, though. If you work with people who don't want to put the effort in, or who want to monopolise the conversation, or have a really bad attitude (like the kind of attitude I tell you to avoid in the next section), it's not going to be so much fun.

Instead, make sure you work with people who are ready to work together, who can talk without fighting and who have a positive, can-do (or if-we-can't-do-then-maybe-we-can-work-it-out-together) attitude. I promise you, it'll be better than slogging through it on your own.

Getting Your Head On Straight

If you were playing a sport, you'd make sure you were in good physical shape before you started. You wouldn't play hockey with a broken arm or hobble onto the basketball court on crutches, I hope.

Studying maths is much the same, only it's more of a mental workout than a physical one. You need to make sure you're in good shape mentally, or else you'll underperform (and perhaps do yourself an injury – believe me, you don't want to end up with a twisted mind!)

In this section, I give you some ideas about how to make sure you're at peak mental fitness before you start studying.

Please do not confuse this with Brain Gym. We do not speak of Brain Gym.

Sorting out your attitude

It's OK, it's OK, it's OK: I'm not accusing you of having an attitude. Sheesh. What I mean is that you ought to have an attitude if you want to do well at A level (and I presume you do). As a tutor, I've noticed several things about the attitudes of students who tend to succeed and the ones who struggle. (**Please note:** This is a personal observation rather than anything I've done a rigorous study on.) The students who do poorly say things like the following:

- ✔ 'We haven't been taught that.'
- ✔ 'I don't like [topic].'
- ✔ 'I'll just hope for an easy paper.'

That's an attitude, all right: it's passive, and it's based on hope and preference rather than effort. 'We haven't been taught that' is particularly telling – it says 'Education is something that happens to me' rather than 'I take responsibility for my learning.' Instead of these horrible phrases, the students who do well say things like this:

- ✔ 'I've not seen that before. Can you point me at some resources?'
- ✔ 'How can I avoid mistakes on [topic]?'
- ✔ 'What's the nastiest thing they can ask about [topic]?'

See the difference? If you want to succeed, you need to *engage*. Figure out what you need to know, and make sure you learn it. Your teachers are there to help you do that, but they can't learn it for you.

Talking yourself up

Explain to me something, would you, while I tug at my cloth cap and wave my walking stick in the air? Why do young people today come out of exams and say, 'That went *terribly!* I don't think I got a single question right!' when they know perfectly well they're going to get an A? I'm sure they did it in my day, too, but I never bothered to challenge them on it.

Do you do that? Not necessarily the 'going to get an A' bit, although I hope that's the case. If you do talk like that, then stop it. It's silly and unhelpful: if you keep telling yourself you're bad at maths or that you're not good in exams, you'll end up making yourself believe it – which just leads to stress and misery. Stress and misery? Those are bad things best avoided.

Instead, you should be saying good things after exams: 'I'm pretty sure I got that question right' or 'I got lost on question 8, but I think I've picked up a few marks.' You're a decent mathematician, or they wouldn't be letting you do A level – own that!

This goes for everything in maths – if you say, 'I can't do this', you're reinforcing that idea. Simply changing the statement to 'I haven't figured this out yet' or even to 'I'm missing a piece here. I wonder what it is' changes your perception: instead of a statement of static helplessness, you've turned it into a dynamic statement in which you're moving forwards and looking for a solution. It sounds a bit daft, but it makes a *big* difference to your outlook.

Don't say things to yourself that you wouldn't say to a friend who was struggling. You wouldn't tell your best friend she was useless, so don't tell yourself that, either. You'd encourage her to keep at it and reassure her it'll come good in the end, and she'd feel a bit better. That works when you do it to yourself, too.

Coping when things go wrong

Things *do* go wrong when you're studying. You will make mistakes. You will find things difficult. *I* make mistakes. *I* find things difficult. *Stephen Hawking* makes mistakes and finds things difficult. It's normal. It's part of the process.

The main difference between a good mathematician and a mediocre one is how they handle mistakes, bad test scores, tellings-off and other general disappointments. There's a lot of talk among teachers and politicians about 'resilience', and this is what they mean: it's the ability to bounce back when things go wrong.

I think everyone has a day when they wake up and maths has suddenly become *really* hard. For some people, it's algebra; for others, calculus; for others, logarithms. For me, it was functions at university – I remember turning the paper over in the exam and realising that I had no idea on the first two questions and just about recognised some of the words in the third. Yes, I even checked to make sure I was sitting the right exam. Proper, full-blown panic attack in the exam, and only just about got on top of it in time to answer enough questions to get me through with a mediocre grade.

Here's how I handled it:

- ✔ **Take deep breaths.** If you're not breathing properly, your brain isn't working properly. Put everything down for a few seconds, try to clear your brain and breathe deeply. (I was later taught to count to seven on the in-breaths and to eleven on the way out, but just making sure you fill your lungs up is a good start.)

- ✔ **Sit up straight.** This helps with the breathing – and it also helps you feel more confident. If you behave like a confident person, sometimes you can borrow some confidence!

- ✔ **Talk yourself up.** I mention this earlier in the chapter, but saying, 'Oh my God, I can't do this, I'm going to *fail*, aargh!' isn't especially helpful. Instead, say things like 'I *was* paying attention' and 'There *is* something on here I can do' and 'Let me try this and see how far I get.' After you get going, you can often fool your brain into keeping going.

- ✔ **Change things around.** In that exam, I knew I was toast for the first couple of questions, so I started with the last question and worked backwards through the paper. By the time I'd calmed down and got up a head of steam, I'd figured out what the third question was asking.

If you get regular panic attacks, *please* go and see a doctor. I suffered from undiagnosed mental health problems for many years and would really rather you didn't. They're horrible.

Setting Up a Study Plan

By and large, you can expect your school classes to lead you through the content you need in a fairly logical order and at a pace that means you'll cover everything in time for the exam. And that's fine, although I've worked with many students whose teachers fell ill at a critical stage and were not able to cover everything, some students who missed school themselves and ended up far behind, and a few students whose schools messed up and were not able to cover the whole course. It can happen.

The key to avoiding these calamities is to take charge. It's your A level. Figure out what you need to know and when you need to know it by, and make sure you learn it. Not only will it give you a fallback in case things go pear-shaped at school, but when *you* decide what you're going to work on, it's much less of a chore to do it.

Being your own pacemaker

There is nothing – *nothing* – in the rules that says you have to go at the same pace as your class. Ideally, nothing your teacher teaches you in class should be a surprise: if you've read ahead in the textbook to familiarise yourself with what's in the module, you're much better-placed to understand what you're being told.

This is part of what I mean by 'taking responsibility for your learning'. The more active you are about deciding what to learn and when to learn it, the more effective you'll be at learning it! Rather than simply doing your homework and saying, 'I'm done,' it's worth taking ten minutes to see what's on the menu for the next class.

You may not be used to learning from a textbook rather than from a teacher, but it's a skill that'll serve you well, both in other courses and at university (should you choose to go there). Here are some tips on how to do it:

- ✔ **Read the text.** This might sound like an obvious thing, but course textbooks aren't just a source of homework questions. They also give plenty of description of what a topic is about, why it's important and what it's used for. So if another student tries to derail the class by saying, 'But when will we ever use *this*?' you can say, 'Well, it's a really big thing in physics' or similar and shut them up.

- ✔ **Make notes.** As you go through, write down anything you think is important. Put things into your own words, if you can. Again, it's all about active learning: if you're translating something from Mathematician into English, you're thinking about it rather than letting it wash over you.

- ✔ **Come up with questions.** Take special care over anything that's not 100 per cent obvious to you, and think about what questions you could ask to put it right. The more specific questions you can ask, the better help you'll get. 'I don't get logarithms' is a much harder thing to fix than 'I tried this question, and here's what I did. I suspect the mistake is around here, but I can't see it.'

Finding the topics

Deciding which topics to revise isn't quite as big a job as doing the actual revision, although for some people it seems to take just about as long! There are four main ways to find topics to work on:

✔ **Ask your teacher for a list.** This is the laziest option of the lot, but it's worth a punt – some teachers have ready-made revision lists to hand out. If not, you need to do some legwork yourself.

✔ **Look at your textbook.** If your textbook is the exam-board produced one, you can expect each chapter of the book to correspond roughly to one question in the exam. It doesn't always work that way, but it's a decent guideline.

✔ **Look at past papers.** Skim over the questions. Any that look frightening? They're probably a good place to start.

✔ **Look at the syllabus.** Not for the faint-hearted, this one: go to the exam board's website and download the syllabus. You'll get a long, slightly scary document – but it tells you everything they expect you to know and how complicated you can expect the questions to be. (It's aimed at teachers rather than students, so there's a good deal of jargon, but in terms of finding topics, it's hard to think of a more comprehensive source.)

Using spaced learning

Just about everyone who's ever revised for an exam has crammed for an exam. Unfortunately, the evidence is that cramming is of limited use. You can get a nice warm glow and a feeling of familiarity with the material, but if you want to be able to recall the material, spaced learning is a much more effective way to work.

If you revisit material after a few minutes (at the end of a session, perhaps), then after a few hours (the next morning), then after a few days, and then after a few weeks, you can cement ideas in your mind.

Mark on a planner or calendar when you expect to read and review each topic – organisation pays off!

Assessing yourself

How can you tell how well (or not well) you're doing? Have you really got to grips with everything you've been studying, or are there subtle gaps in your knowledge?

My best advice is not to wait for your mocks to start doing past papers. The A level hasn't changed an awful lot in the last ten to fifteen years, for much of which there were two exam diets a year, so there are usually twenty past papers or so to choose from. If you run out, there are Solomon papers and other practice tests – there's no shortage of exams for you to try.

If you get hold of a paper, work through it under exam conditions and then grade it against the mark scheme, you're probably putting yourself ahead of most A level students; however, you can wring more reward out of your effort with just a few extra hacks:

- **Rewrite correct answers for anything you got wrong.** This is a chore, but it's really helpful for when you're revising – and it also gets you into the habit of writing down the right thing. You can also use the mark scheme to find out what kind of language they're after, especially in 'explain' questions – you'll often see the kinds of words and ideas that will get you marks in the future.

- **Make notes on which topics to review.** If you've dropped marks out of carelessness, that's something you can fix by taking more care. If you've dropped marks because you've forgotten how to do something, it's worth looking back at the relevant chapters in this book and/or your textbook – and doing examples until you find it easy.

- **Swap marking duties with a friend.** One way to get a real feel for what it's like to be on the other side of the table is to mark someone else's work. You can start to see the kinds of things that annoy examiners and what makes their lives easier. Then you can start avoiding the annoying things and do the nice things, which will translate into extra marks.

- **Give each other feedback.** Related to this, if there's something you understand and your buddy doesn't, go over it with him or her. Helping someone with a problem really cements it in your own mind (and that person may be able to help you in the same way with something else).

- **Look for alternative methods.** The mark scheme often offers several ways of doing a problem. Have a look at these to see if there's anything you can learn.

Quick-Fire Revision Techniques

Just like different students work better in different environments, different students respond better to different revision techniques. To round off the chapter, here's a selection of some approaches I've seen students use effectively. Feel free to pick whichever ones seem most likely to work for you – or better yet, experiment.

If you have another approach that works for you (especially game shows I might have missed), I'd love to hear about it! I'm @icecolbeveridge on Twitter, or you can reach me through my website at www.flyingcoloursmaths.co.uk.

Calendar of crosses

When Jerry Seinfeld was starting out as a comic, he suffered from a lack of material. He set himself a challenge: to write something every day. He bought a cheap calendar, and every day he wrote something, he'd put a cross on the calendar.

After a while, he noticed that the crosses were forming a nice chain pattern, and he didn't want to break the chain. He kept writing and writing, and I gather he's done pretty nicely for himself.

Multimillionaire mathematicians are about as rare as multimillionaire stand-up comics, but it's rare to see a mathematician working as a barista (except out of choice). However, you can steal Seinfeld's trick to make sure you do a bit of maths work every day, whether it's rewriting some notes to make them clearer, working through an exercise or reading your textbook.

Mark the days you've worked, and don't break the chain!

Game shows

Game shows are one of my guilty pleasures – it's a rare evening that doesn't find me catching up with my *Pointless* friends. Adding an element of competition to knowing things makes knowing them so much more interesting, which is why I recommend working with a friend or friends in a quiz show format. Some ideas:

- ✔ Get a big black chair and ask each other quick-fire questions à la *Mastermind*. You can even pick specialist subjects for the first round!

- ✔ If one of your friends (possibly you) is significantly stronger than the others, play *The Chase*: give the weaker student a head start, and see if he or she can answer enough questions to get home before the chaser catches up.

- ✔ For a change, you can try *Jeopardy!*: the quizmaster gives an answer, and the contestants need to figure out the question. For example, the quizmaster might say, 'The quadratic has two real roots when the discriminant is this,' to which the question may be 'What is positive?'

Pretty much any quiz show (or game) you enjoy can be turned into a mathematically themed one. It's certainly less dull than systematically answering questions on paper.

Cheat sheets

Although spaced learning is a great way to embed knowledge in that brain of yours, you can still *use* your short-term memory to help you in an exam. My favourite short-term memory hack is to make up a small sheet of things to remember – things you often forget or get wrong – and read through it in the last few minutes before the exam. When you get into the exam, write down the things you've just remembered – and then they're there for you to refer to at leisure.

You're obviously not allowed to bring your notes into the exam, so please don't try. If you're caught, and there's a good chance you would be, you can expect draconian punishment. (I've heard stories – possibly legends – of students having all their qualifications taken away after being caught cheating, including GCSEs. It's really not worth the risk.)

If you make your cheat sheet colourful and engaging, remembering what's on it will be much easier. Practise looking through it for a few minutes and then writing down the contents from your short-term memory. You need to be confident you can get everything down.

Sticky notes everywhere

I once had a student who, in an effort to memorise the definitions and techniques she needed, wrote them out on sticky notes and attached them to every door in her house. Before going through a door, she had to answer the question on the sticky note or give the correct definition. As soon as she was confident on a particular note, she'd replace it with something else she needed to know.

I don't think it's a coincidence that the student who built maths revision into her day-to-day routine ended up with an A*. Do you?

Chapter 3

All the Algebra You Missed

. . . .

In This Chapter

▶ Understanding operations with Boodles

▶ Working with powers

▶ Expanding and simplifying

▶ Handling fractions

▶ Solving simultaneous equations

. . . .

*L*isten, I'm not saying you *did* sneak through GCSE with shaky algebra skills. I'm just saying that some people *do* – and those people tend to find A level really tough from the get-go. If quadratics make you queasy and fractions drive you frantic, now would be the ideal time to get on top of them before you find yourself getting crushed by them.

Things like solving simple equations and simultaneous equations, dealing with brackets (expanding and factorising), and knowing a few squares and cubes are also taken for granted at A level.

Luckily, once you have the hang of them, they're easy marks come the exam!

The Brilliance of Boodles: Understanding the Order of Operations

There is an order to mathematics. I like to think of it as a wall, with bricks of knowledge supporting more complicated bricks, which support further bricks in turn. For example, the operations of arithmetic have a clear complexity structure.

The first real sums you learn to do are adding and subtracting. You then move on to multiplying and dividing – which are really shortcuts for adding or subtracting the same things over and over again. Sometime after that, you get into powers and roots, which are shortcuts for repeated multiplication and (to a lesser extent) division.

It's not a coincidence that the order of operation laws (which you might know as BIDMAS or BODMAS or PEMDAS or similar) move down from the most complicated calculations (*indices* – or, for normal humans, powers and roots) to the more fundamental sums (adding and subtracting).

You can see a visual representation of this hierarchy in Figure 3-1. You can use this little table to understand the order of operations, your power laws (see the next section) *and* the logarithm laws you see in Chapter 5.

Brackets

Powers   Roots

Figure 3-1: Multiplication   Division
The Boodles
hierarchy.
Addition   Subtraction

Illustration by John Wiley & Sons, Inc.

Incidentally, I prefer this to BIDMAS as a way to remember the order of operations – it clearly shows that multiplication and division are on the same level; if you're not careful about how you learn BIDMAS, you find yourself always dividing before you multiply, which isn't quite right. Instead, the hierarchy shows that these two operations have the same precedence, which means you can do them in any order. Try it! Multiply 12 by 7 and divide it by 3; you get the same answer as if you divide 12 by 3 and multiply by 7.

This property – being able to choose the order in which you do operations on the same level – is extremely useful for complicated calculations: working out $\frac{9\times8\times7\times6}{4\times3\times2\times1}$ is much less complicated if you treat it as $(9\div3)\times(8\div4)\times7\times(6\div2)=3\times2\times7\times3=6\times21=126$ rather than $3,024\div24$.

How do you remember this hierarchy? I know of two ways: one is by thinking of my favourite restaurant in Bozeman, Montana, which was called Boodles. It was a 'Posh Restaurant that / Made Delicious / Apple Sauce'. You can, of course, come up with your own mnemonic – it's probably better for you if you do. Alternatively, you can spot the word 'AMP' going up the left-hand side of the table and fill in the opposite operations on the right.

Adding and subtracting are two sides of the same coin: each of them undoes the other. You can even think of subtracting a positive number as adding a negative number. Because they're really the same thing, they have the same precedence and can be done in whatever order suits you best. In a similar way, division is the same thing as multiplying by a fraction, so multiplication and division have the same precedence; also, taking a root is the same thing as raising a number to a fractional power, which is why powers and roots are on the same level.

Practising Your Power Laws

The traditional way of learning your power laws is to look at powers of ten. For example, you know that $10^2 = 100$ and $10^3 = 1,000$; you also know $100 \times 1,000 = 100,000$, which is 10^5 – so you can immediately remember that when you multiply two powers with the same base, you have to add the powers. Similarly, you can figure out that dividing two powers with the same base means you take away the powers.

Taking it a step further, you might notice that $\left(10^2\right)^3 = 100 \times 100 \times 100 = 1,000,000$, or 10^6. If you raise one power to another, you have to multiply the powers together. If you're very good at the opposites game, you may spot that taking a root must be something to do with fractional powers.

Luckily, I have a mnemonic device that makes working all this stuff out a bit less haphazard. In this section, I show you the power of Boodles, list some values to memorize and tell you how to make quick work of fractional powers.

Boodles power

The Boodles hierarchy (see the preceding section) offers a tidy way to recall the power laws. For example, remember that multiplying powers with the same base means adding the power $\left(x^a x^b = x^{a+b}\right)$. Look back at the Boodles table in Figure 3-1: that means going down a row. Raising a power to another power means you'd then multiply the powers – which also means going down a row. Very straightforward, right? The table also shows you that taking, for example, the fourth root of a number is the same as raising it to a $\frac{1}{4}$ power – effectively *dividing* the power by 4.

Here are some other facts you may want to keep in mind:

- ✔ Anything to the power of 1 is itself.
- ✔ Anything to the power of 0 (except 0 itself) is 1; 0^0 is not defined, and maths usually pretends it's not a thing.

    ✔ 1 to the power of anything is 1.

    ✔ 0 to the power of anything (except 0) is 0.

    ✔ A negative power gives a reciprocal. For example, $4^{-2} = \dfrac{1}{4^2} = \dfrac{1}{16}$.

In Chapter 5, I show you how to put all of this together to work out something like $\left(64x^6\right)^{-2/3}$.

Knowing your squares, cubes and powers

Knowing your square numbers up to about 20^2 isn't necessary for GCSE or for A level. After all, if you can multiply two numbers together, you can square anything you like. However, knowing your squares makes life significantly easier, especially when you're taking square roots. The same goes for cubes up to 10^3 and powers of 2 and 3 up to 2^6 and 3^6. I suggest putting these down on note cards and getting them learnt – it'll save you a lot of headaches in the long term.

Squares

$1^2 = 1$	$6^2 = 36$	$11^2 = 121$	$16^2 = 256$
$2^2 = 4$	$7^2 = 49$	$12^2 = 144$	$17^2 = 289$
$3^2 = 9$	$8^2 = 64$	$13^2 = 169$	$18^2 = 324$
$4^2 = 16$	$9^2 = 81$	$14^2 = 196$	$19^2 = 361$
$5^2 = 25$	$10^2 = 100$	$15^2 = 225$	$20^2 = 400$

Cubes

$1^3 = 1$	$6^3 = 216$
$2^3 = 8$	$7^3 = 343$
$3^3 = 27$	$8^3 = 512$
$4^3 = 64$	$9^3 = 729$
$5^3 = 125$	$10^3 = 1,000$

Powers of 2

$2^1 = 2$

$2^2 = 4$

$2^3 = 8$

$2^4 = 16$

$2^5 = 32$

$2^6 = 64$

Powers of 3

$3^1 = 3$

$3^2 = 9$

$3^3 = 27$

$3^4 = 81$

$3^5 = 243$

$3^6 = 729$

Handling nasty fractional powers

If you were asked to work out $8^{\frac{10}{3}}$, you'd be forgiven for thinking, 'I know! I'll find the 10th power of 8 and then cube root it.'

Good luck with both of those things, at least without a calculator: $8^{10} = 1,073,741,824$. I doubt I could work that out on paper in under ten minutes – and finding its cube root would take longer still.

Much simpler: take the root before you do the power. The cube root of 8 is 2, and $2^{10} = 1,024$. This ties into the general principle of making things simpler before you make them more complicated.

Another, even more elegant approach is to spot that $8 = 2^3$, so $8^{\frac{10}{3}} = \left(2^3\right)^{\frac{10}{3}} = 2^{10}$ directly.

Expanding Brackets and Simplifying

You've probably been multiplying brackets out since year 8 or so, but in case it's passed you by, it's one of the most fundamental skills you need to be comfortable with at A level. In this section, I offer you a few tips for staying organised and avoiding mistakes.

Using the basic grid

If you tend to use FOIL or crab claws, I'm afraid I need to convert you to the grid method of expanding brackets. That's not just blind prejudice; there's a reason for it: FOIL and crab claws are absolutely fine as long as you're dealing with brackets with two terms in them, such as $(x+2)(2x-3)$. However, if you have something like $(x+2)(2x^2-3x+1)$, you're a bit stuck. And you often do have something like that, I'm afraid.

Here's how the grid method works on each of those examples – follow along in Figure 3-2:

1. **Draw a grid with as many rows as you have terms in the first bracket and as many columns as you have terms in the second bracket.**

   In the first example, it's a 2-by-2 grid; in the second, it's 2-by-3.

2. **Write the terms of the first bracket to the left of the first column, one term per row.**

   In both cases, the first row will have an x to the left, and the second row will be preceded by a 2.

3. **Write the terms of the second bracket above the top row, one term per column.**

   In the first example, the top row reads $2x$, -3 (see Figure 3-2a); in the second, it reads $2x^2$, $-3x$, 1. (see Figure 3-2c).

4. **In each of the squares, multiply the thing at the top of its column by the thing on the extreme left.**

   In the first example, the top-left square is $2x \times x = 2x^2$; below that is $2x \times 2 = 4x$; the top-right square is $-3 \times x = -3x$, and the bottom-right is $-3 \times 2 = -6$. (See Figure 3-2b.)

   In the second example, the first column has $2x^3$ and $4x^2$; the next column is $-3x^2$ and $-6x$; finally, the last column is x and 2. (See Figure 3-2d.)

5. **Add up everything in the squares you've just worked out, simplifying where you can.**

   The first one becomes $2x^2 + 4x - 3x - 6 = 2x^2 + x - 6$. The second one is $2x^3 + 4x^2 - 3x^2 - 6x + x + 2 = 2x^3 + x^2 - 5x + 2$.

You can extend this method to any number of terms by making grids that are as big as you need them to be (although you're unlikely to see anything even as complicated as three terms multiplied by three terms).

(a)

	$2x$	-3
x		
2		

(b)

	$2x$	-3
x	$2x^2$	$-3x$
2	$4x$	-6

Figure 3-2: Two multiplication grids.

(c)

	$2x^2$	$-3x$	1
x			
2			

(d)

	$2x^2$	$-3x$	1
x	$2x^3$	$-3x^2$	x
2	$4x^2$	$-6x$	2

Illustration by John Wiley & Sons, Inc.

Squaring things

One of the commonest 'silly mistakes' I see in students' work is for them to write something like $\left(x + 2 \right)^2 = x^2 + 4$.

Noooooo!

No. No, no, no. No, no, no, no, *NO!* Don't do that. Please. $(x+2)^2$ is most certainly not equivalent to x^2+4. (Check it with, say, $x=1$). It's the same thing as this:

$$(x+2)(x+2)$$
$$= x^2 + 2x + 2x + 4$$
$$= x^2 + 4x + 4$$

Every time you miss the middle x term out of a squared bracket, a kitten dies. Don't do it. Won't somebody think of the kittens?

Expanding several brackets

Expanding several brackets is a pain in the whatsit. If all the brackets are the same – for example, $(2x-3)^5$ – then you want the binomial expansion from Chapter 8; otherwise, you need to take things a step at a time. Imagine you have to expand $(2x+3)(x-4)(x+1)$. Here's what to do:

1. **Check you actually have to expand it.**

   If you're trying to sketch a curve, it's already in the best possible form. If you're trying to differentiate it and you don't know the product rule, then I'm afraid you do have to expand.

2. **Pick the trickiest pair of brackets to expand, and expand them.**

   Here, I'd say $(2x+3)$ and $(x-4)$ are the hardest. Multiplying them gives you $(2x^2-5x-12)$. See Figure 3-3a.

3. **Multiply this by the next-hardest bracket.**

   I get $2x^3 - 3x^2 - 17x - 12$. See Figure 3-3b.

4. **If you have more brackets, keep going until you've used them all.**

   In this case, you're done.

5. **Look at the first and last terms to check whether your answer makes some sort of sense.**

   I've multiplied a $2x$, an x and another x to get $2x^3$, and I've multiplied 3, -4 and 1 to get -12, both of which are right.

   This check doesn't guarantee the middle bits are correct, but if it's wrong, you know you've made a mistake somewhere.

Figure 3-3:
Multiplying
out with the
grid method.

(a)

	x	-4
$2x$	$2x^2$	$-8x$
3	$3x$	-12

(b)

	$2x^2$	$-5x$	-12
x	$2x^3$	$-5x^2$	$-12x$
1	$2x^2$	$-5x$	-12

Illustration by John Wiley & Sons, Inc.

Fiddling About with Fractions

If someone came to me straight after their GCSE and said, 'What one thing should I do over the summer to give me the best chance at A level?', I would almost certainly say, 'Get good at fractions.' It's the area I've seen most students struggle with. 'I don't like fractions' is about as common a grumble in my classroom as 'I'm quite tired'.

So, once and for all, let's get the rules of fractions sorted out, starting with numbers and moving on to algebra afterwards. There aren't many rules, and most of them involve trying to turn a nasty fraction into something nicer.

Manipulating fractions with numbers

Repeat after me: a fraction just means 'divide the top by the bottom'. For example, $\frac{1}{2}$ is exactly the same thing as $1 \div 2$. Look, the sign for *divide by* even looks like a fraction! In this section, I take you through the rules and try to explain why they work the way they do.

But before I do that, I need to get a soapbox out. Here we go. Is this thing – *skrreeeek!* – on? Right: *Fractions are better than decimals*. You might have spent your entire GCSE turning things into decimals and writing abominations like $1.5x$ instead of $\frac{3}{2}x$, but that stops now. The only times you should give an answer as a decimal are when the question uses decimals or when you're told to give answers to a certain number of decimal places.

Fractions are, in general, a more efficient way of representing numbers than decimals, and they're *far* easier to multiply: $\frac{1}{7} \times \frac{2}{3} = \frac{2}{21}$, but $0.\dot{1}4285\dot{7} \times 0.\dot{6} = \ldots$ well, I don't know. I'd have to work it out from the fraction (somewhere a bit less than 0.1, maybe 0.095?).

Relatedly, you'll see a lot of questions in A level that ask for an *exact* answer. You might think that means 'write down all the digits on your calculator display'.

If you do that, the examiners will laugh at you and mark you down as one of *those people*. An exact answer usually involves a root, a power of e, a logarithm and/or π. It's just how mathematicians write – it's much more efficient to write down e^{π} than it is to write down 23.1406926328. . . – and it would still be more efficient even if the decimal expansion didn't go on forever.

The last piece of mathematical style advice I have for you is to avoid mixed fractions such as $3\frac{1}{3}$ in favour of top-heavy fractions such as $\frac{10}{3}$. When you're multiplying and dividing, that sort of fraction is easier to work with than a mixed fraction is.

Equivalent fractions

A half $\left(\frac{1}{2}\right)$ is the same thing as $\frac{50}{100}$ or $\frac{7}{14}$ or $\frac{0.994}{1.988}$ or infinitely many other fractions – all of which have a bottom that's double the top. That's the hallmark of equivalent fractions: The ratio between the top and the bottom is the same. Although all these fractions have the same value, the *canonical* (usually best) form is $\frac{1}{2}$ – it uses whole numbers and has no common factors to the top and bottom. This is called the *lowest form* of a fraction, and it's how you should almost always leave things. (The only exception? If you're explicitly told you need not simplify your answer.)

This means you can simplify fractions whenever you have a common factor on the top and the bottom.

You also need to be able to convert between mixed fractions (such as $2\frac{1}{2}$) and their more useful top-heavy (*improper*) counterparts. In this case, the top-heavy version of $2\frac{1}{2}$ is $\frac{5}{2}$. Here's how I worked it out:

1. **Convert the leading number (the bit that isn't a fraction) into a fraction.**

   You keep the same base as the given fractional part, and the top is the original number multiplied by the bottom. Here, 2 becomes $\frac{4}{2}$.

2. **Add this value to the given fractional part.**

   $$\frac{4}{2} + \frac{1}{2} = \frac{5}{2}$$

3. **Check it makes sense.**

   $\frac{5}{2}$ is between 2 and 3, so it's not obviously wrong.

To go the other way (although the only reason you would ever do that would be if the question asked for it explicitly), here's what you'd do with $\frac{60}{7}$:

1. **Divide the top by the bottom, and make a note of the whole number part of the answer and the remainder.**

   Here, you have 8, remainder 4.

2. **The whole number part goes at the front; the remainder goes over what you divided by to begin with.**

   You get $8\frac{4}{7}$.

3. **Again, check your answer is plausible.**

   8 is $\frac{56}{7}$ and 9 is $\frac{63}{7}$, so you're in the right ballpark.

If you've done the last recipe correctly, you should feel a small twinge of distress deep in your soul. Mixed fractions are evil and should make you feel slightly dirty.

Multiplying fractions

Multiplying a number by a fraction is about the simplest thing you can do with fractions. You multiply the top by the number and leave the bottom alone:

$$3 \times \frac{4}{7} = \frac{12}{7}$$

The reason this works is that $3 \times (4 \div 7)$ is the same thing as $(3 \times 4) \div 7$. Because multiplication and division are on the same level of the Boodles hierarchy from a few sections back, you can regroup them like this with no consequence.

To multiply a fraction by a fraction isn't much harder: You multiply the tops together, and you multiply the bottoms together: $\frac{5}{7} \times \frac{5}{6} = \frac{25}{42}$. Again, the reason lies in the Boodles hierarchy: because you can multiply and divide in whatever order you like, $(5 \div 7) \times (5 \div 6) = (5 \times 5) \div 7 \div 6 = (5 \times 5) \div (7 \times 6)$.

Watch out: you can use this trick only on things that are on the same Boodles level; $\frac{7}{4+6}$ is certainly not the same thing as $\frac{7}{4} + \frac{7}{6}$!

Powers and roots of fractions

Raising a fraction to a power isn't really difficult. The long way of doing it is to multiply the fraction by itself over and over again; after you've done that a few times, you notice that all you've done is raise the top to the given power and raise the bottom to the given power. For example,

$$\left(\frac{3}{4}\right)^4 = \left(\frac{3}{4}\right)\left(\frac{3}{4}\right)\left(\frac{3}{4}\right)\left(\frac{3}{4}\right) = \frac{3^4}{4^4} = \frac{81}{256}.$$

The same idea goes for roots (after all, they're just powers): $\sqrt[3]{\frac{64}{125}} = \frac{\sqrt[3]{64}}{\sqrt[3]{125}} = \frac{4}{5}$.

Again, you apply the root to the top and to the bottom, and you're done!

Be careful of negative numbers – you're not allowed to take square roots (or any even roots) of these, and you will break maths if you try. Look ahead to Chapter 4 for details!

Dividing fractions

Dividing fractions is a little more complicated than multiplying but not much: flipping the second fraction (strictly, finding its reciprocal) turns the division into multiplication. For example, $\frac{3}{4} \div \frac{2}{5} = \frac{3}{4} \times \frac{5}{2} = \frac{15}{8}$.

You can convince yourself of this with a bit of algebra. If you call your answer x, then $\frac{3}{4} \div \frac{2}{5} = x$. Multiply both sides by $\frac{2}{5}$ to get $\frac{3}{4} = \frac{2}{5}x$. To get rid of the $\frac{2}{5}$, you can multiply both sides by $\frac{5}{2}$ to get $\frac{3}{4} \times \frac{5}{2} = x$.

Adding and subtracting fractions

The last fraction trick you need to know about is probably the one you learnt first at school (some things just get taught the wrong way round). Adding fractions with the same bottom is easy. You just add the tops and leave the bottoms alone: $\frac{1}{7} + \frac{3}{7} = \frac{4}{7}$.

In fact, you can think of the bottom as a unit of measurement, so a seventh is like a gram or a kilometre – and you'd never try to add inches to millimetres directly; you'd always convert one into the other (or into some common system) first. That's precisely what you do with adding or subtracting fractions: you find equivalent fractions with the same bottoms. For example, suppose you have to work out $\frac{1}{5} + \frac{3}{4}$. The most convenient unit to work with is 20ths – the 20 comes from 5×4. A fifth is the same as $\frac{4}{20}$ (multiplying the top and bottom by 4), while three-quarters is the same as $\frac{15}{20}$ (multiplying the top and bottom by 5) – and now you can add them together to get $\frac{19}{20}$.

Here's a recipe:

1. **Multiply both the top and bottom of the first fraction by the bottom of the second. Write this down somewhere.**

   This converts the first fraction to a better 'unit'.

2. Multiply both the top and bottom of the second fraction by the bottom of the first. Write this down next to the answer from Step 1.

This converts the second fraction to the same unit as the first.

3. Now add or subtract the tops, and write your answer over the common bottom.

And you're done.

Ninja tricks

Before you go on to algebraic fractions (spoiler: all the rules are just the same), here are a few ninja tricks that aren't strictly necessary to working with fractions, but they can really cut the time you spend on questions.

Ninja Trick Number 1 is the 'cross-cancel', a trick you can use when multiplying fractions. For instance, if you're working on $\frac{8}{15} \times \frac{3}{4}$, the average student blindly works it out – correctly – as $\frac{24}{60}$ before cancelling it down to $\frac{2}{5}$. The ninja, though, spots that 8 and 4 have a common factor of 4, which means you can divide the top of one fraction by 4 and the bottom of the other by 4 to get $\frac{2}{15} \times \frac{3}{1}$. Similarly, there's a 3 common to the 3 and the 15, so you can simplify further to $\frac{2}{5} \times \frac{1}{1}$, which is obviously $\frac{2}{5}$.

Ninja Trick Number 2 is called the *sausage rule*. Usually, I'm not a great fan of tricks that have only one application, but the fraction = fraction pattern is common enough that it's worth knowing. (If you've read *Basic Maths For Dummies* – hello, thanks for sticking with it! – you might recognise this as the Table of Joy). The trick works if you have a fraction equal to another fraction and an unknown somewhere in it – such as $\frac{5}{x} = \frac{3}{7}$. Of course, you may solve this using regular algebra (multiply both sides by x to get $5 = \frac{3}{7}x$; multiply both sides by 7 to get $35 = 3x$; divide both sides by 3 to get $x = \frac{35}{3}$); alternatively, you can write down the answer with the sausage rule. Here's how:

1. Draw a sausage around the numbers 'neighbouring' your unknown – the 7 to the right and the 5 above it. Multiply these numbers together.

Here, you'd do $5 \times 7 = 35$.

2. Divide your answer by the other number.

$\frac{35}{3}$ is your final answer for x.

This trick is especially useful for things like the sine rule, which is in Chapter 4.

Algebraic fractions

There's some Very Good News about algebraic fractions: they follow exactly the same rules as regular fractions. You multiply them, divide them and raise them to powers in exactly the same way. You add them and subtract them the same way. You even simplify them the same way, by dividing the top and the bottom by a common factor.

For example, to multiply two algebraic fractions, such as $\frac{x-3}{x+2} \times \frac{x-2}{x-3}$, you multiply the tops together to get $\frac{(x-3)(x-2)}{(x+2)(x-3)} = \frac{x-2}{x+2}$, because there's a common factor of $x-3$ on the top and bottom.

To divide two fractions, you flip the second one and multiply:
$$\frac{x-2}{x^2-9} \div \frac{x^2-4}{x-3} = \frac{x-2}{x^2-9} \times \frac{x-3}{x^2-4} = \frac{(x-2)(x-3)}{(x^2-9)(x^2-4)}, \text{ which simplifies further:}$$
each factor on the bottom is the difference of two squares, which you can write as $(x-3)(x+3)$ and $(x-2)(x+2)$. Then $\frac{(x-2)(x-3)}{(x-3)(x+3)(x-2)(x+2)}$ simplifies to $\frac{1}{(x+3)(x+2)}$. There's no need to multiply the brackets on the bottom out. It's much easier to see what's going on with the bottom factorised.

Adding and taking away fractions with algebra are also similar. To simplify $\frac{2}{x} + \frac{3x+1}{x-2}$, just find the common denominator, $x(x-2)$, and convert both fractions into the correct form: $\frac{2(x-2)+x(3x+1)}{x(x-2)} = \frac{3x^2+3x-4}{x(x-2)}$. If the top could be factorised, you would factorise it.

When taking away fractions, make sure the top of the second one goes in a bracket – losing a minus sign is one of the most frustrating mistakes you can make!

Solving Single Equations

Solving algebraic equations should be pretty much second nature to you from GCSE. You need to be able to solve linear equations in one variable (*linear* means 'looking like a straight line' – either or both sides of the equation will be of the form $ax+b$) and quadratic equations (where either or both sides look like ax^2+bx+c).

Basic linear algebra

The golden rule for solving equations is that whatever you do to one side, you must also do to the other. If you add 2 to the left side of an equation, you add 2 to the right. If you divide one side of the equation by 7, you divide the other side by 7, too.

If, for example, you have $2 - 3x = 4x - 44$, here's what you do. (***Note:*** In what follows, I mean 'lowest' in the mathematical sense: –3 is lower than 2.)

1. **Take the lowest number of *x*s away from both sides.**

   Here, subtract $-3x$ (or rather, add $3x$) on each side to get $2 = 7x - 44$.

2. **Take away the number on the same side as the remaining *x*s.**

   Here, subtract -44 (or add 44) on each side to get $46 = 7x$.

3. **Divide both sides by the number of *x*s.**

   $$\frac{46}{7} = x$$

Doing the operations in this order means you do as little work as possible. Other orders are available.

Dealing with simple fractions

When you've got an unknown on the bottom of a fraction in an equation, it's usually a good idea to get it onto the top. The way to do that is to multiply both sides by the ugly, unknown-containing denominator. It's quite often a good idea to do that even if you have a constant on the bottom of a fraction – however good you are at fractions, there's less scope for error when you're working with whole numbers.

For example, if you have $\frac{8}{3 - x} = 4x$, you should multiply both sides by $(3 - x)$. That gives you $8 = 4x(3 - x)$, or $4x^2 - 12x + 8 = 0$, a quadratic you can then solve. Using one of the methods in the upcoming "Solving quadratics" section, you should get $x = 1$ or $x = 2$.

If you have something like $\frac{x}{3} = 2 + 7x$, you could try to do it all with fractions... or you could multiply the whole thing by 3 to get $x = 6 + 21x$, turning it into a linear equation you can solve ($-6 = 20x$, so $x = -\frac{3}{10}$).

At A level, fractions are better than decimals for giving answers. Here, –0.3 isn't a wrong answer; it's just not what the examiners really want to see.

Doing rougher rearrangement

One of the nastier GCSE tricks you need is rearranging things like $p = \dfrac{x+3}{x-p}$ to get x in terms of p. Here's the form:

1. **Get everything on the top line by multiplying both sides by the bottom of the fraction.**

   You get

   $$p(x-p) = x+3$$

2. **Expand the brackets and rearrange so all of the xs (or whatever you happen to be solving for) are on the same side.**

   $$px - p^2 = x+3$$
   $$px - x = p^2 + 3$$

3. **Factorise as much as possible.**

   $$x(p-1) = p^2 + 3$$

4. **Lastly, divide by the bracket to get x on its own.**

   $$x = \frac{p^2 + 3}{(p-1)}$$

Solving quadratics

There are three sensible methods (that I know of) for solving a quadratic equation:

- ✔ Factorising (elegant, but it doesn't always work)
- ✔ Completing the square (mathematically sound, but it's fiddly)
- ✔ Using the formula (reliable, but it requires good knowledge of squares in Core 1)

I cover completing the square in Chapter 6. Right now, I take you through the quadratic formula (so you have a reliable method) and, after that, factorising (so you have an elegant one).

The quadratic formula

Given a quadratic equation that isn't easy to factorise (something along the lines of $4x^2 + 3x + 7 = x^2 - 9x + 20$), here's how to solve for x:

1. **Arrange it in the form $ax^2 + bx + c = 0$.**

   You get $3x^2 + 12x - 13 = 0$.

2. **Write down your a, b and c.**

   Here, $a = 3$, $b = 12$ and $c = -13$.

3. **Use the quadratic formula: $x = \dfrac{-b \pm \sqrt{b^2 - 4ac}}{2a}$.**

   This gives you $x = \dfrac{-12 \pm \sqrt{144 - 4(3)(-13)}}{6} = \dfrac{-12 \pm \sqrt{300}}{6}$.

4. **Simplify your answer.**

   $x = -2 \pm \dfrac{5\sqrt{3}}{3}$ or $\dfrac{-6 \pm 5\sqrt{3}}{3}$ if you prefer.

You're not given the quadratic formula. You need to remember $x = \dfrac{-b \pm \sqrt{b^2 - 4ac}}{2a}$ or learn to complete the square!

The most common problems students have with the quadratic formula are forgetting the $\pm$ and, especially when b is negative, not squaring it properly. Remember, whatever real number b is, its square can't be negative.

Factorising (with just x^2)

Before you start factorising, think about $(x + a)(x + b)$. If you expand that, you get $x^2 + (a + b)x + ab$. If you add the numbers-that-aren't-x together, you get the coefficient of x in the quadratic; if you multiply them, you get the constant term at the end.

That's what factorising boils down to: finding the a and b that work.

If you're asked to factorise $x^2 - 2x - 15$, you need to find two numbers that multiply together to make -15 but add to make -2. Luckily, -15 doesn't have too many factors: they could be -1 and 15 (which add to 14), -3 and 5 (which add to 2), 3 and -5 (which add to -2) or 1 and -15 (which add to -14). Clearly, you want 3 and -5 as your a and b – either way around, it doesn't matter. This quadratic factorises as $(x + 3)(x - 5)$.

Once more with a recipe, this time for factorising $x^2 - 13x + 36$:

1. **List the factor pairs of the last number.**

   The factors of 36 are 1 and 36, 2 and 18, 3 and 12, 4 and 9, 6 and 6, and the negatives of all of those.

2. **Add up each pair and find which one adds up to the middle number.**

   9 and 4 make 13, which suggests (correctly) that –9 and –4 make –13.

3. **Write down these magic numbers after *xs* in brackets.**

   $$(x-9)(x-4)$$

Factorising quadratics (with ax^2)

Now, I have a little confession to make. I didn't know how to factorise quadratics with anything in front of the x^2 until one of my early students showed me how. At that point, I had two maths A levels, a degree in maths, a PhD in maths and four years of postdoctoral research experience – which goes to show you can get quite a long way with significant gaps in your knowledge. (For the record, I tended to work around the problem using the quadratic formula.)

You, on the other hand, should get the hang of it. It's easy once you know how. Here's how to factorise $6x^2 - 5x - 6$:

1. **Start by multiplying the number of x^2s by the number at the end.**

   Here, you get –36.

2. **List the factor pairs of this number.**

   You could have –1 and 36, –2 and 18, –3 and 12, –4 and 9, –6 and 6, or the same things with the signs reversed.

3. **Add up each pair and find what adds up to the middle number to get your magic numbers.**

   Here, 4 and –9 would do the trick.

4. **Split the middle term up into the two magic numbers multiplied by x.**

   $$6x^2 + 4x - 9x - 6$$

5. **Factorise the first two terms and the last two terms, separately.**

   $$2x(3x+2) - 3(3x+2)$$

6. The bracket is now a common factor, so you can combine the whole expression into a pair of brackets.

$$(2x-3)(3x+2)$$

It doesn't matter which way around you write the magic numbers in Step 4 – if you had written $6x^2 - 9x + 4x - 6$, you would have factorised it as $3x(2x-3) + 2(2x-3) = (3x+2)(2x-3)$, which is also correct.

This last step can also be done with a multiplication grid, as in Figure 3-4.

Figure 3-4:
Factorising
with a multi-
plication
grid.

(a)

$6x^2$	$-9x$
$+4x$	-6

(b)

	$2x$	-3
$3x$	$6x^2$	$-9x$
2	$+4x$	-6

Illustration by John Wiley & Sons, Inc.

Solving Simultaneous Equations

When you have two (or more) equations that need to be satisfied at the same time, they're called *simultaneous equations*. (*Simultaneous* means 'at the same time'. Maths is quite logical sometimes.) At A level, you're sometimes explicitly told to solve them (as in, the question says, 'Solve these simultaneous equations'), and sometimes you have to use simultaneous equations without being told (for example, you need to find where two curves cross or you just simply have two equations that both need to be true).

There are two sensible ways to solve simultaneous equations: substitution and elimination. In this section, I show you how both of those work with linear and nonlinear simultaneous equations.

Linear simultaneous equations

Linear simultaneous equations are ones where both equations are – go on, guess – linear! That means if you drew them out, you'd get straight lines. (That's the case if neither variable is squared, you're not dividing by either variable, and you're not multiplying the variables together.)

Which method is easier depends both on you and on the question. Both methods ought to work in all cases, but sometimes one is much less work than the other. As a general rule, if you have $y =$ something or $x =$ something, substitution is probably less hassle.

Substitution method

The substitution method does exactly what it says on the tin: you rearrange one equation, if necessary, to get one of the variables on its own and then substitute into the other equation. For example, if you were solving

$$x + y = 1$$
$$5x + 3y = -13$$

you might use this recipe:

1. **Pick a variable to substitute for.**

   Suppose you pick y.

2. **Rearrange one of the equations to get your variable on its own.**

   Here, you might get $y = 1 - x$.

3. **In the other equation, replace every instance of your variable with what it's equal to, in a bracket.**

   Here, you have $5x + 3(1 - x) = -13$.

4. **Solve that.**

$$5x + 3 - 3x = -13$$
$$2x = -16$$
$$x = -8$$

5. **Work out the other variable using the value you just worked out.**

$$y = 1 - x = 1 - (-8) = 9$$

So for this example, the substitution method gives you $x = -8$ and $y = 9$.

Elimination method

Generations of students have been taught to add and subtract simultaneous equations from each other, and generations of students have messed up their minus signs and ended up with wrong answers.

If you know what you're doing with that method, go ahead and do it with my blessing. However, I'm not going to teach that to you here; instead, here's a more sensible variation, using the same example as before ($x + y = 1$ and $5x + 3y = -13$).

1. **Pick a variable to eliminate and get a multiple of that variable on its own.**

   For the sake of variety, I'll pick x. In the first equation, I get

   $$x = 1 - y$$

   And in the other, I get

   $$5x = -13 - 3y$$

2. **Get the same number of your variable in each equation.**

   Here, I multiply the first equation by 5 to get

   $$5x = 5 - 5y$$

   In general, you need to multiply both equations by something to get the coefficients to be the same. The number and the sign need to be the same – multiply one of the equations by –1 if they aren't.

3. **Set the right-hand sides equal to each other.**

   You know that $5x$ is the same as $5 - 5y$ and also the same as $-13 - 3y$, so

   $$5 - 5y = -13 - 3y$$

4. **Solve.**

   $2y = 18$, so $y = 9$.

5. **Go back and find the other variable.**

   $$x = 1 - y = 1 - 9 = -8$$

The answer is $x = -8$ and $y = 9$.

Nonlinear simultaneous equations

Nonlinear simultaneous equations are ones where at least one of the equations is – you guessed it – not a linear equation. It might be a quadratic, it might be a circle, it might be something else altogether. (The other is, thankfully, *usually* a linear equation.)

In this kind of situation, there's often more than one possible answer (in principle, there could be infinitely many solutions; $y = \sin(x)$ and $y = 0$ cross over and over again, forever in both directions), but you won't be asked for more than a few solutions. The most I've ever seen in an A level question is six.

As with linear simultaneous equations, there are two main methods for solving nonlinear ones: substitution and elimination. However, I'm only going to show you substitution, on the grounds that it's almost always simpler.

As an example, I'll take the hardest C1 simultaneous equations question I've ever seen, although numbers have been changed to protect the author from copyright violation claims:

$$x + y + 1 = 0$$
$$x^2 + xy + y^2 = 3$$

1. **Using the linear equation, get one of the variables on its own.**

   $x = -1 - y$ will do nicely.

2. **In the other equation, replace every instance of that variable with its equivalent (in a bracket).**

   $$\left(-1-y\right)^2 + \left(-1-y\right)y + y^2 = 3$$

3. **Expand, simplify and solve.**

   $$\left(1 + 2y + y^2\right) + \left(-y - y^2\right) + y^2 = 3$$
   $$y^2 + y - 2 = 0$$
   $$\left(y-1\right)\left(y+2\right) = 0$$
   $$y = 1 \ \text{ or } \ y = -2$$

4. **Go back to your linear equation and solve for the other variable.**

   If $y = 1$, $x = -1 - 1 = -2$. And if $y = -2$, $x = -1 - \left(-2\right) = 1$.

Then write down the solutions, clearly marking which pairs go together: here, the answers are $\left(-2, 1\right)$ and $\left(1, -2\right)$.

Chapter 4

Shaping Up to Graphs and Shapes

. . . .

. . . .

*Y*ou spend quite a lot of your GCSE studies contemplating graphs: bar graphs, pie charts, histograms, box plots, cumulative frequency, stem-and-leaf diagrams (a completely made-up graph, by the way), frequency polygons (almost completely made up), velocity–time graphs . . . the list goes on. You need precisely none of those in the core part of your A level – although some of them will crop up in statistics, if you take that option, and velocity–time graphs are used in mechanics.

In core maths, the only graphs you care about are the graphs of functions – straight line graphs, quadratic graphs, trigonometric graphs. You'll extend that to cubic graphs and polynomials in general, to circles and to more complicated functions in A level – and you'll be working out things like tangents, normals and the shapes they make – but in this chapter, the only graphs you care about are the three I mentioned.

You do a lot with shapes at GCSE, too, especially triangles and circles. This chapter gives you a chance to review those skills.

Circle Theorems

If you listen to my *Wrong, But Useful* podcast or read my blog, you'll know that I have a thing about circle theorems, which boils down to this: I was a proper, card-carrying maths/physics researcher for the better part of a decade. I studied the sun, which – you've probably noticed – is *round*. If there's one area of maths you would expect circle theorems to come in useful, it's looking at the sun.

But no. In all those years, the only time I needed a circle theorem was when a cartoonist friend had drawn a circle around a yoghurt pot and then realised he didn't know where the centre was. I told him to use the diameter theorem, twice, and see where the diameters crossed.

So circle theorems win the Colin Award for the least useful bit of compulsory maths in GCSE, fighting off strong competition from probability questions involving pens. Luckily, you don't need to know many circle theorems for A level; unfortunately, you still need a few, which I outline next.

Diameters

Repeat after me: the angle opposite a diameter is a right angle, as shown in Figure 4-1. There are several ways to show this, although my favourite is to note that angles A, B and C sum to 180° because of the big triangle, while $C = A + B$ because the two small triangles are both isosceles. That means $2C = 180°$, making C a right angle.

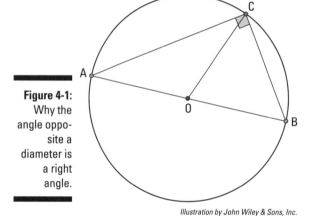

Figure 4-1:
Why the angle oppo-site a diameter is a right angle.

Illustration by John Wiley & Sons, Inc.

Tangents

A tangent to a circle is perpendicular to its radius. One way to show this is by contradiction, as in Figure 4-2a: if the angle between the radius (AB) and tangent (BT) weren't 90°, then the line that *is* perpendicular to the radius (BC) would have to be a chord cutting the circle somewhere else (C). The radius to that point (AC), the original radius (AB) and the chord (BC) form an isosceles triangle, but it would have two right angles in it – which is clearly impossible, as the angle at the centre (BAC) would have to be 0°. The original assumption that the angle between the radius (*AB*) and the tangent (*BT*) wasn't 90° must have been mistaken.

The other tangent fact you need to know is that two tangents extending to the same point are the same length. You can show this with Pythagoras: looking at Figure 4-2b, you can see that Δ*OAP* and Δ*OBP* are both right-angled (at *A* and *B*, respectively), both have the same hypotenuse (*OP*), and both have a short side of the same length (because *OA* and *OB* are both radii). That means the final pair of sides must be the same length, so the tangents *AP* and *BP* are equal.

(a)

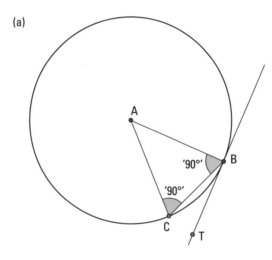

(b)

Figure 4-2:
(a) Why a tangent is perpendicular to the radius;
(b) why tangents that extend to meet at the same point are equal.

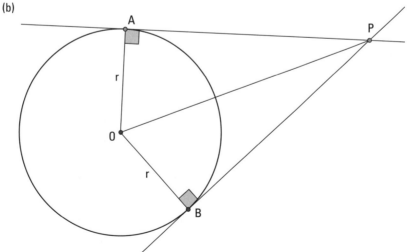

Illustration by John Wiley & Sons, Inc.

Chords

You need to know two things about *chords*, lines that go across a circle but not necessarily through the centre:

✔ Connecting the points where a chord intersects the circle to the centre gives you an isosceles triangle (because the radii have the same length).

✔ As a result, the perpendicular bisector of the chord goes through the centre of the circle.

Figure 4-3 shows you these concepts in action.

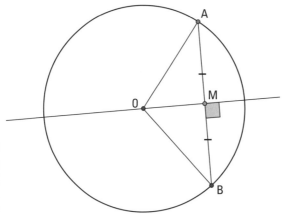

Figure 4-3:
Chord
theorems.

Illustration by John Wiley & Sons, Inc.

Straight Lines

I know, I know. You're probably bored with $y = mx + c$. It's been drilled into you so many times over the last few years. If you're confident on straight line graphs, feel free to skip over this section. However, if you need a refresher, you're in the right place!

At A level, you come across other ways of writing a straight line – they contain the same information, and the other ways have their own advantages (which I extol in Chapter 10), but $y = mx + c$ has the key ideas you need.

To define a straight line algebraically, you need two things:

✔ **A gradient, *m*:** How steep the line is

✔ **A point on the line:** Typically, where the line crosses the *y*-axis, from which you can work out *c*

You need to understand the link between the graph of a straight line and its equation: a point lies on a line if and only if its coordinates make the equation 'work'. For instance, if your line is $y = 3x - 5$, the point $(2, 1)$ lies on the line because when you replace y with 1 and x with 2, the left-hand side of the equation gives you 1 and the right-hand side also gives you 1. The point $(0, 3)$ is *not* on the line, because substituting in the coordinates gives you 3 on the left and –5 on the right, which aren't equal.

Sketching roughly

It's not necessarily obvious how you measure the steepness of a line, even when you know it's how far up you go for every unit you go to the right (the rise over the run). As far as a sketch goes, here's a rough guideline:

- ✔ If your gradient is bigger than 1, your line should be steeper than 45° (up and to the right).
- ✔ If your gradient is 1, your line should be roughly 45° to the axes (up and to the right).
- ✔ If your gradient is between 0 and 1, your line should make an angle less than 45° with the x-axis (still going up and to the right).
- ✔ If your gradient is 0, it should be parallel to the x-axis.
- ✔ If your gradient is negative, the same rules apply, except your line should be going down and to the right.

One of the most important principles of sketching graphs at A level is that your graph doesn't have to be perfect. It just needs to have the right sort of shape. So if you're sketching a straight line, you need to make sure it's straightish, it's roughly steep enough, and it crosses the y-axis in roughly the right place. You don't need to mark out numbers on your axes, and you don't need to measure anything.

Getting a gradient

One other thing about straight lines for now: how to find the gradient from the equation. The recipe is really short but usually a bit fiddlier than it looks. Imagine you have the equation $4y = 7 - x$:

1. If your equation isn't in the form $y = mx + c$, rearrange it so it is.

Here, you put the $-x$ in front of the +7 and divide it all by 4, like this:

$$y = -\frac{1}{4}x + \frac{7}{4}$$

2. The gradient is the number in front of the x.

That's $-\frac{1}{4}$ here.

3. The y-intercept – where the line crosses the y-axis – is the other number.

Here, it's $\frac{7}{4}$.

In Chapter 10, you see how to find the equation of the line starting from two points or from a point and a gradient.

Beware of the difference between 'the coordinates of a point' and 'the y-intercept'. The coordinates of the point where this line crosses the y-axis are $\left(0, \frac{7}{4}\right)$; the y-intercept is the single number, $\frac{7}{4}$.

Other Graphs

Again, Chapter 10 has more details on how to understand and sketch graphs, but there are a few things you should know from GCSE.

The first is an extension of what you know about the equation of a line: a point lies on a curve if and only if its coordinates make the equation work. For example, if you have $y = 3x^3 - 3$, you can tell that the point $(2, 21)$ is on the curve because putting $x = 2$ and $y = 21$ into the equation gives you 21 on the left-hand side and 21 on the right-hand side – the equation is true. By contrast, $(0, 0)$ isn't on the graph, because substituting in gives you 0 on the left and –3 on the right – the sides aren't equal.

You can plot the graph of any function (at least, the functions you'll see at A level) by working out the y-value that goes with any value of x and putting the points on a graph.

The second thing to know is that curves generally don't have a fixed gradient – it changes as you go along. You can't use the same rules of thumb as you do for the gradient of a straight line – you'll need to use techniques of differentiation, which are in Chapters 14, 15 and 17.

You're supposed to have a passing knowledge of what a quadratic graph looks like (see Figure 4-4), whether it's the right way up (such as $y = x^2 + 3x + 2$) or upside-down (like $y = 3 - x^2$). You should be able to identify reciprocal graphs (such as $y = \dfrac{1}{x}$) and simple cubics (like $y = x^3$; see Figure 4-5). Lastly, knowing what the graphs of sine and cosine look like is a real help – they look like Figure 4-6 (note that the x-values are in radians, a glorious way of measuring angles that I cover in Chapter 11).

(a)

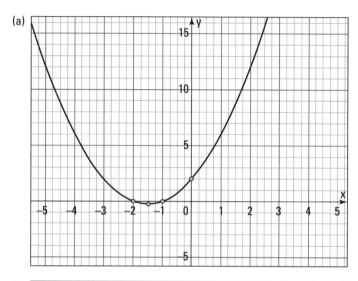

(b)

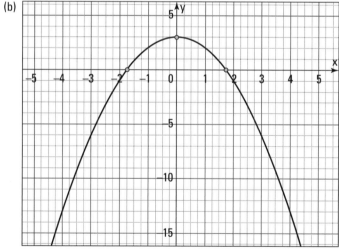

Figure 4-4:
Quadratic
graphs.

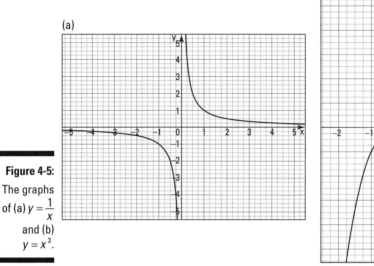

Figure 4-5: The graphs of (a) $y = \dfrac{1}{x}$ and (b) $y = x^3$.

Trigonometry

The triangle is the best shape in all of maths. It's the simplest of the polygons, yet it's incredibly versatile. It's also tremendously useful – any time you want to do maths with coordinates or geometry in general, the chances are there's a triangle behind what you're doing.

You won't be surprised to discover, then, that basic trigonometry – SOH CAH TOA and Pythagoras's theorem – are assumed knowledge for Core 1 and Core 2. You tend to relearn the sine rule and cosine rule for Core 2, but I include them here because you should have seen them before.

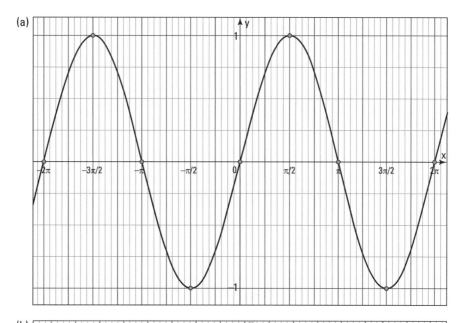

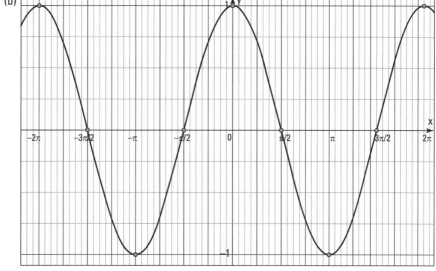

Figure 4-6:
The
graphs of
(a) $y = \sin(x)$
and (b)
$y = \cos(x)$.

Pythagoras's theorem

In the great tradition of maths naming conventions, Pythagoras's theorem wasn't discovered by Pythagoras – it was known to the Babylonians some centuries before. As far as I can gather, Pythagoras just had a better publicist than his Babylonian predecessor. The theorem is usually written as

$$a^2 + b^2 = c^2$$

Although I prefer the more descriptive version:

$$\text{opp}^2 + \text{adj}^2 = \text{hyp}^2$$

A few quick things to note:

- ✔ Pythagoras works only for right-angled triangles. If you're not sure you have a right angle, you should use the cosine rule.

- ✔ If Pythagoras's theorem holds true for three sides of a triangle, then the triangle is right-angled.

- ✔ A common error with Pythagoras is adding the squared hypotenuse to one of the short sides squared instead of taking away (that's why it's important to write it out with *opp*s and *hyp*s). Another is to forget to square-root your answer.

Sanity checks on Pythagoras are always a good idea. Is your hypotenuse still the longest side? Is it less than the other two sides added together? If not, then I'm afraid you've made an impossible triangle – either that, or a mistake. If you've made a mistake, smile and fix it!

SOH CAH TOA

Being asked to solve right-angled triangles is unusual (at least in the Core modules – if you're doing Mechanics, they crop up all the time). However, SOH CAH TOA is assumed knowledge, and it's occasionally quite handy.

Figure 4-7 shows a right-angled triangle and the ratios associated with it.

Remember that 'opposite' and 'adjacent' are relative to the acute angle in question – if you decide to work with the other angle, you need to reconsider what's opposite to it and adjacent to it!

For example, if you know the longest side of a right-angled triangle (the hypotenuse) is 15 cm long and one of the other angles is 23°, you can find either of the remaining sides as follows. Let's pick the adjacent side. Why? Just cos.

Figure 4-7:
A right-angled triangle and the associated ratios.

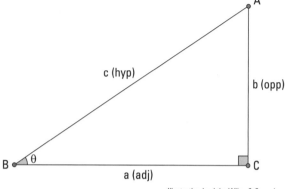

$$\sin(\theta) = \frac{\text{opp}}{\text{hyp}}$$

$$\cos(\theta) = \frac{\text{adj}}{\text{hyp}}$$

$$\tan(\theta) = \frac{\text{opp}}{\text{adj}}$$

Illustration by John Wiley & Sons, Inc.

1. **Pick the ratio that involves the two things you know and the one thing you want to know.**

   Here, that's $\cos(\theta) = \dfrac{\text{adj}}{\text{hyp}}$.

2. **Substitute in the values you know.**

   $$\cos(23°) = \frac{\text{adj}}{15}$$

3. **Rearrange to get your unknown on its own.**

   $$\text{adj} = 15\cos(23°)$$

4. **Work this out on the calculator.**

   To two decimal places, it's 13.81 cm. (In later chapters, I extol the virtues of exact answers where possible; when giving measurements, answers as rounded decimals are generally fine unless you're told otherwise.)

5. **Check the number makes sense.**

   It does – it's shorter than the hypotenuse but not too short.

 In this chapter, I use degrees to measure angles, like you probably have all through your career. In a later chapter (Chapter 11), I explain in great detail why degrees are dangerous, wrong-headed and harmful, but for now, you can keep your calculator in 'degrees' mode. *For now.*

Sine rule

You use the sine and cosine rules when you have a triangle that doesn't have a right angle – although they work perfectly well on triangles that do.

You always start by setting up your triangle with labels like the one in Figure 4-8: the angles are marked *A*, *B* and *C*, and the side opposite each angle is marked with the same letter in lower-case.

Figure 4-8:
A triangle
labelled and
ready to use
with the
sine and
cosine
rules.

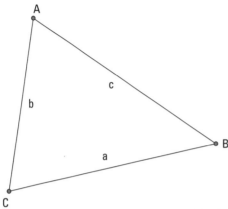

Illustration by John Wiley & Sons, Inc.

You're not given the sine rule in the formula book, but it's not too difficult to remember:

$$\frac{a}{\sin(A)} = \frac{b}{\sin(B)} = \frac{c}{\sin(C)}$$

The sine rule works just as well the other way up, but this way up has a geometrical meaning! If you draw a circle through the three points of the triangle, $\frac{a}{\sin(A)}$ and its friends are all equal to the circle's diameter.

You use the sine rule when you know a 'pair' – that is, you know an angle and the length of the opposite side, two things represented by the same letter (one capital, one not). For example, suppose you know an angle is 45°, and the side opposite it has a length of 21 cm. Another angle is 70°; what is the length of the side opposite that?

Here's what you'd do:

1. **Draw a picture.**

   Even if you already have one, drawing a picture is a good habit. Yours should look like the one in Figure 4-9.

2. **Label the angles and sides.**

   I'd call *A* 45°, *B* 70°, *a* 21 cm and *b* the-thing-you-want-to-know.

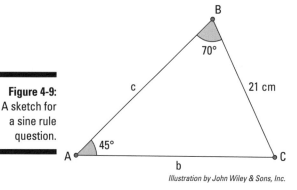

Figure 4-9:
A sketch for
a sine rule
question.

Illustration by John Wiley & Sons, Inc.

3. Write out the sine rule, replacing letters with their values.

$$\frac{21}{\sin(45°)} = \frac{b}{\sin(70°)}$$

(You can ignore the third fraction because you don't care about any of the other values.)

4. Solve for the missing variable.

$$b = \frac{21\sin(70°)}{\sin(45°)} \approx 27.91 \text{ cm}$$

If you're solving for an angle, don't forget to use your $\sin^{-1}$ button to get rid of the sine.

This recipe can mislead you: when you're finding a missing angle using the sine rule, there is an *ambiguous case*. The question will either tell you whether you're looking for an acute or obtuse angle or make it clear in some other way which one makes sense.

Cosine rule

You use the cosine rule when you care about all three sides of a triangle – if you have two sides and the angle between them (a sort of elbow shape) and want the final side, or if you have all three sides and want an angle – as in Figure 4-10. (That second case? That's the reverse cosine rule, which is in the next section.)

For all of the boards except WJEC, the formula book very kindly gives you the cosine rule:

$$a^2 = b^2 + c^2 - 2bc\cos(A)$$

(a) 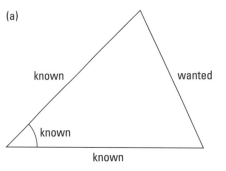 (b)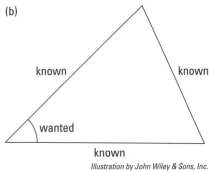

Figure 4-10:
Situations in
which you
use the
cosine rule.

Illustration by John Wiley & Sons, Inc.

Here's how you'd find the missing side of the triangle on the left of Figure 4-10, with sides of 12.9 cm and 15.2 cm and an angle between them of 40°.

1. **Label your triangle. The angle you know is *A*, and each side has a lower-case letter opposite the corresponding capital letter.**

   It doesn't matter which is *B* or *C* in this case – I'm picking $b = 12.9$ and $c = 15.2$, but you can try it the other way around if you don't believe me.

2. **Replace all the known letters with the corresponding numbers.**

   $$a^2 = 12.9^2 + 15.2^2 - 2(12.9)(15.2)\cos(40°)$$

3. **Type this into your calculator.**

   You should get 97.038 or so.

4. **Square-root this to get *a*.**

   To two decimal places, it's 9.85 cm.

5. **Check it makes sense.**

   Forty degrees is a fairly small angle, so you'd expect the side opposite it to be similar to or smaller than the other sides. It looks about right.

The most common mistakes in a cosine rule question are leaving out Step 4 (forgetting to square-root the answer from Step 3) and making calculator errors in Step 2. The number of people who work out $(b^2 + c^2 - 2bc)\cos(A)$ is shocking. Don't be one of them.

Reverse cosine rule

You can also use the cosine rule to find an angle if you know all three sides. This concept comes with a 'take care' warning – it's an easy one to mess up, so I offer you two recipes, and you should pick the one that works best for you. Both of them rely on rearranging the formula to get *A* on its own, but they differ in when you do the rearranging!

Imagine you've got a triangle with sides of 5, 7 and 9 cm, and you want to know the largest angle – which, if you think about it for a moment, is opposite the longest side. Here's the 'algebra first' way to find it:

1. Draw and label the triangle, as in Figure 4-11.

Call the angle you're looking for A, and label the others in appropriate pairs.

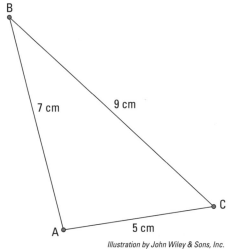

Figure 4-11:
Cosine rule
triangle.

Illustration by John Wiley & Sons, Inc.

2. Write down the cosine rule.

$$a^2 = b^2 + c^2 - 2bc \cos(A)$$

3. Rearrange to get the a on the right and the A on the left.

$$2bc \cos(A) = b^2 + c^2 - a^2$$

4. Divide everything by $2bc$.

$$\cos(A) = \frac{b^2 + c^2 - a^2}{2bc}$$

5. Substitute in your lengths.

$$\cos(A) = \frac{5^2 + 7^2 - 9^2}{2(5)(7)} = -\frac{7}{70} = -\frac{1}{10}$$

6. Type in $\cos^{-1}(Ans)$, which works out the inverse cosine of the result in Step 5.

You get about 95.7°.

If you're someone who loves formulas, the one at the end of Step 4 may be one to add to your learning list. If you're not, don't worry; working it out on the fly isn't too time-consuming if your algebra is good.

If you don't like the algebra method, I advise doing some work on your algebra – not out of nastiness, but because you'll need strong algebra skills to do well at A level. However, the cosine rule is one of the places where you can sometimes scrape along without it.

1. **Draw and label the triangle, as in Figure 4-11.**

   Call the angle you're looking for A, and label the others in appropriate pairs.

2. **Write down the cosine rule.**

   $$a^2 = b^2 + c^2 - 2bc\cos(A)$$

3. **Fill in your numbers.**

   $$81 = 25 + 49 - 70\cos(A)$$

4. **Simplify.**

   $$7 = -70\cos(A), \text{ so } \cos(A) = -\frac{1}{10}.$$

5. **Take the inverse cosine to work out A.**

   $$\cos^{-1}(Ans) \approx 95.7°$$

Although this approach works well with nice round numbers like we have here, if you've got nastier numbers, such as the answers to previous parts of the question, you'll want to use the *Ans* button cleverly or store the answers in your calculator.

Storing numbers in your calculator

Waiting for a class in Thame, Oxfordshire, I settled down in the café with a fine piece of literature: not Shakespeare, not Austen, not Hardy, but the manual for the Casio FX-83 GT PLUS calculator. (You can find it as a free download on the Internet if you search for it; it's riveting. Five stars.)

One of the things I learned while digesting it (along with a carrot cake and a cappuccino, of course) was how to store numbers in the calculator's memory – which is really useful if you don't want to have to type them in repeatedly. What you do is

1. **Work out your number.**

2. **Press SHIFT, then RCL, then a button with a red letter above it.** The screen should say something like '3359->X'. That's your number stored in memory.

To use the number, simply press the red alpha button followed by the letter you want. If you clear your calculator and type 'X=', (or whichever letter you chose), you'll get the number you worked out. '2X=' will print out double the number. If you store a, b and c as A, B and C (sadly, you don't have the option of using lower-case letters, which would be even better), you can type the cosine rule in directly, which is why I mention it.

Areas

A good general approach to areas (and volumes) is to split shapes up into smaller shapes you know about. In its most extreme form, that's what integration is – adding up a bunch of tiny areas and volumes to find a total.

In this section, I take you through some of the area rules you might remember from GCSE.

Triangles

The area of a triangle is $\frac{1}{2}bh$ – in my experience, it's one of the best-remembered formulas for A level students (although some persistently muddle it up with Pythagoras). Not everyone knows where it comes from, though. Every triangle is half of a parallelogram, and every parallelogram has the same area as a rectangle with the same base and perpendicular height.

Perhaps a bit less-well-remembered is the trigonometric version: the area is $\frac{1}{2}ab\sin(C)$, using the same labelling conventions as the earlier 'Trigonometry' section – C is the angle between sides a and b. Why does the formula work? Well, if you look at Figure 4-12, you see that the height of a triangle labelled like this works out to be $b\sin(C)$, so $\frac{1}{2}ab\sin(C)$ is the same thing as $\frac{1}{2}bh$!

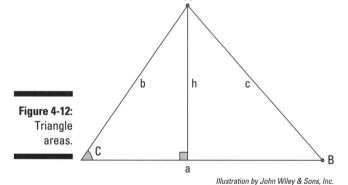

Figure 4-12:
Triangle
areas.

Illustration by John Wiley & Sons, Inc.

Sectors, arcs and segments

A *sector* is a slice of a circle like the one in Figure 4-13a, made of two radii and a bit of the circumference – an *arc*. If you're in a sharp mood, you'll notice there are two ways you could go around a circle – the smaller sector is called the *minor* sector, and the larger one is the *major* sector.

You can use GCSE techniques to find the length of the arc and the area of the sector if you know the angle at the centre and the radius: if your angle is $A°$, you have $\frac{A}{360}$ of a circle, so the arc length is $\frac{A}{360} \times 2\pi r$ and the area is $\frac{A}{360} \times \pi r^2$.

Similarly, you can find the area of a *segment*, which is the part of a circle chopped off by a chord. Again, the bigger piece is the *major* segment, and the smaller one, the *minor* segment. If you draw radii to the ends of the chord, as in Figure 4-13b, you can see that the segment is the difference between a sector and a triangle – and you know how to find their areas! The area of a segment is $\frac{A}{360} \times \pi r^2 - \frac{1}{2} r^2 \sin(A)$. In Chapter 11, you get to use a different measure of angle (the radian), which makes the formula much nicer.

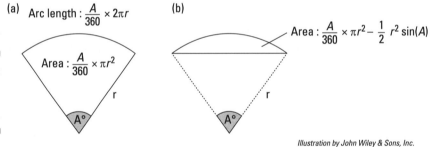

(a) Arc length : $\frac{A}{360} \times 2\pi r$

(b) Area : $\frac{A}{360} \times \pi r^2 - \frac{1}{2} r^2 \sin(A)$

Area : $\frac{A}{360} \times \pi r^2$

r

$A°$

Figure 4-13: Areas of a sector and a segment.

Illustration by John Wiley & Sons, Inc.

Part II
Arithmetic and Algebra

Binomial Expansion: $(9 - 8x)^{-1/2}$

1	$-\dfrac{1}{2}$	$\dfrac{3}{8}$	$-\dfrac{5}{16}$
$\dfrac{1}{3}$	$\dfrac{1}{27}$	$\dfrac{1}{243}$	$\dfrac{1}{2,187}$
1	$-8x$	$64x^2$	$-512x^3$
$\dfrac{1}{3}$	$\dfrac{4}{27}x$	$\dfrac{8}{81}x^2$	$\dfrac{160}{2,187}x^3$

The laws for powers and logarithms are usually just given to you — and that's fine, if you're good at remembering things. However, you may prefer to understand why they are what they are; if you do, pop over to www.dummies.com/extras/asalevelmathsuk for an explanation.

In this part . . .

- ✔ Pick out powers, surds and logarithms.
- ✔ Practise polynomials.
- ✔ Find factors, remainders and fractions.
- ✔ Sort out sequences and series.
- ✔ Make sense of functions.

Chapter 5

With Great Power Comes...

*I*f you know what you're doing with powers and surds, A level maths gets a whole lot easier. I often get feedback from students along the lines of 'Why should we have to work with surds when my calculator does that for me?' which I half-understand but now disagree with, for several reasons. (I'm sure, when *my* old A level maths teachers read this, they'll gleefully write in to remind me that I asked them that exact question in the late 1990s.)

Firstly, you need to be able to deal with surds in Core 1, where you don't have a calculator. That's not an excuse to forget surds as soon as you're out of that exam, though; you also need to be able to manipulate things like $\sqrt{x}$ and $y^{7/5}$ throughout the whole of the course, and your calculator isn't especially helpful with algebra.

After you've got the hang of surds and powers, it's a short step to the logarithm function (a sort of 'undo' for powers) and to working with the most important constant in A level maths – not π, as you might have expected, but e, Euler's constant, which is about 2.72.

Making Sense of Surds

If you ask, 'What's the square root of 3?' the average student will pick up a calculator, tap a few buttons and say, 'It's about 1.732.' This is, of course, true; however, it's not the answer examiners are usually looking for. Examiners want to see $\sqrt{3}$.

Writing the square root of three as $\sqrt{3}$ instead of 1.732. . . has all sorts of advantages (for instance, it takes a long time to write down infinitely many decimal places, and doing sums. . . well, forget about it). As a mathematician, your process should be to work with exact forms (like surds and πs and everything else) for as long as possible and bring the calculator into play only if the question demands it.

Multiplying out

The thing to remember about doing algebra with surds is that they behave just like an x (or any other bit of algebraic furniture), except you know what you get if you square it. For example, if you need to expand $\left(2+3\sqrt{2}\right)\left(3-2\sqrt{2}\right)$, it behaves exactly the same as $\left(2+3x\right)\left(3-2x\right)$, where $x=\sqrt{2}$. You get $6+5x-6x^2$ – only in this case, you know that $x^2=2$, by definition, so $6x^2$ is 12. The final answer is $6+5\sqrt{2}-12$, which simplifies to $-6+5\sqrt{2}$.

Multiplying surds by each other can be a little trickier, but not much. Because surds are well-behaved under multiplication $\left(\sqrt{a}\times\sqrt{b}=\sqrt{ab}\right)$, you sometimes need to be on your toes for things you can simplify. For example, $\sqrt{6}\times\sqrt{15}\times\sqrt{10}=\sqrt{900}=30$.

Rationalising simple denominators

Maths books like to pretend that having square roots on the bottom of a fraction is somehow dirty or wrong (before quite merrily telling you that $\tan\left(30°\right)=\frac{1}{\sqrt{3}}$). It's not – although when the exams tell you to rationalise a denominator, you don't have much choice but to get them off of the bottom and on to the top. There are two kinds of irrational denominators you need to worry about:

✔ Simple ones like $\frac{1}{\sqrt{3}}$

✔ More complicated ones like $\frac{2+\sqrt{2}}{3-\sqrt{2}}$

Dealing with simple ones is, as you'd hope, simple. When you have a single square root and nothing else on the bottom, all you need to do is multiply the top and bottom by the square root, like this:

$$\frac{1}{\sqrt{3}}=\frac{1\times\sqrt{3}}{\sqrt{3}\times\sqrt{3}}=\frac{\sqrt{3}}{3}$$

And there you go! The surd that – for some never-to-be-explained reason – wasn't wanted on the bottom has jumped to the top. Hooray! But what if something like $3 - \sqrt{2}$ is on the bottom? Read on!

Rationalising harder denominators

The trick to rationalising the denominator of a fraction with a bottom of the form $\left(a \pm \sqrt{b}\right)$ is to make use of the difference of two squares. You know that $(x + y)(x - y) = x^2 - y^2$ (or you could work it out by expanding the brackets if you didn't). That's a really nice trick if one of the things on the bottom is a square root: if you multiply (top and bottom) by the same thing but with the opposite sign in between, you get rid of the square root on the bottom. Here's how it works with $\dfrac{2 + \sqrt{2}}{3 - \sqrt{2}}$:

1. **Work out the *conjugate* of the bottom – the same thing with the opposite sign between.**

   Here, it's $3 + \sqrt{2}$.

2. **Multiply top and bottom by the conjugate.**

   The top is $\left(2 + \sqrt{2}\right)\left(3 + \sqrt{2}\right) = 8 + 5\sqrt{2}$. The bottom is $\left(3 - \sqrt{2}\right)\left(3 + \sqrt{2}\right) = 7$.

3. **Put them together as a fraction.**

   The answer is $\dfrac{8 + 5\sqrt{2}}{7}$.

If it's possible to reduce the fraction further, you should. You should also check exactly what the question asks: if you're told to give your answer in the form $a + b\sqrt{c}$, do exactly that – and state what a, b and c are. Make life easy for whoever's marking your work!

Perfecting Powers

Most of the power-related skills you need came up in Chapter 3, which is a cracking read, I strongly recommend it. This section deals with simplifying the nasty sorts of things that get thrown at you.

A very typical Core 1 question gives you a fraction like $\dfrac{6 + 9x^2}{\sqrt{x}}$ and asks you to express it in the form $ax^p + bx^q$. There are several ways to attack this, but here's the one I recommend:

1. **Turn every term in the fraction into a power of x.**

   This would become $\dfrac{6x^0 + 9x^2}{x^{1/2}}$.

   2. Split the fraction into two (or more) fractions.

$$\frac{6x^0}{x^{1/2}} + \frac{9x^2}{x^{1/2}}$$

   3. Apply power rules to the results to make everything powers of x on the top line.

$$6x^{-1/2} + 9x^{3/2}$$

It's much better to write a power as $\frac{3}{2}$ than as $1\frac{1}{2}$ or 1.5. Mixed fractions are the embodiment of pure evil, and decimals... well, I suppose they're okay if you want to be an economist. They're just not okay here.

Learning to Love the Logarithm

When people who are inexplicably proud of their mathematical ignorance come up to me and say, 'I was never any good at maths!' (in the same way they must go up to voice coaches and say, 'I was never any good at talking!'), they usually note that their stumbling block was long division, trigonometry or logarithms. Sometimes it's just because they've mixed up logarithms and algorithms, but usually it's because logs are a bit different to the sums these individuals were used to. There's a whole apparently new set of rules to learn, things start jiggling all over the place, and *who needs them, anyway*?

Well.

Relationships involving powers are critical in physics, biology, economics, chemistry and probably a dozen fields besides (I used them in my PhD thesis). Logarithms turn sums based on those relationships into much simpler calculations. If you're going to do anything scientific (or become an economist), dealing fluently with logarithms will make your life immeasurably easier and will make you look like a god among students. And who doesn't want that?

What are logarithms for?

A logarithm answers the question 'What power do I raise the base to if I want a particular answer?' For example, because $1,000 = 10^3$, $\log_{10}(1,000) = 3$. See the link?

Logarithms were originally used to make multiplication easier – big numbers are much easier to add than they are to multiply, so early 'computers' – people who computed answers – could look up the logarithm of the numbers they wanted to multiply together in a table, add them together and then reverse the calculation to get the product. That was the way things were done until surprisingly recently – log tables were still a fixture in schools when I was born in the late 1970s.

Obviously, you don't use logarithms for tedious sums these days – we have machines for that sort of thing – but logs *are* useful when you have unknown powers. Solving $2^x = 16,384$, for example, is something that can be done without logs (it's 14, by the way), but when you can just type $\log_2(16,384)$ into your calculator and get the answer immediately, why wouldn't you?

Turning powers into logs (and vice versa)

A standard question in Core 2 gives you an equation with an unknown power, such as $3^x = 100$, and asks you to find x to four significant figures. Here's what you do:

1. **Apply logarithms to both sides, using base of whatever's underneath your variable.**

   With 3^x, you use base 3:

   $$\log_3(3^x) = \log_3(100)$$

2. **Use the fact that $\log_a(x)$ and a^x are *inverse functions* to simplify.**

   That just means that $\log_a(a^x) = x$ in the same way that $\sqrt{x^2} = x$. Here,

   $$x = \log_3(100)$$

3. **Work the right-hand side out on your calculator, using either the $\log_\Box\Box$ button or the identity $\log_a(b) \equiv \dfrac{\log_c(b)}{\log_c(a)}$.**

   I get 4.192, to four significant figures.

Some questions may ask for answers to be written in a form using logs, in which case you don't use the calculator at all; you use the laws of logarithms and regular algebra to rearrange things into the required form – but more on that later.

To get rid of a logarithm, you need to raise both sides to the power of the base, which is easier to see in action than it is to explain. If you know $\log_5(x) = 1.2$ and need an answer to four significant figures, then you do the following:

1. **Raise both sides of the equation to the power of the base.**

   In this case, $\log_5(x) = 1.2$ becomes

   $$5^{\log_5(x)} = 5^{1.2}$$

2. Use the fact that $\log_a(x)$ and a^x are inverse functions to simplify.

$$x = 5^{1.2}$$

3. Find this value using your calculator.

Here, it's 6.899, to four significant figures.

Combining and splitting logarithms

Before I do anything else, I'm going to leave Figure 5-1 here for you to have a look at. It shows the order of operations, or Boodles (see Chapter 3).

Brackets

Figure 5-1: Powers   Roots
The order of
operations,
with the log   Multiplication   Division
laws hidden
in it.   Addition   Subtraction

Illustration by John Wiley & Sons, Inc.

Here are the main laws for combining logarithms:

✔ $\log_a\left(x^n\right) \equiv n\log_a(x)$

✔ $\log_a\left(\sqrt[n]{x}\right) \equiv \dfrac{1}{n}\log_a(x)$

✔ $\log_a(xy) \equiv \log_a(x) + \log_a(y)$

✔ $\log_a\left(\dfrac{x}{y}\right) \equiv \log_a(x) - \log_a(y)$

Why have I referenced Figure 5-1? Well, when you're trying to get a logarithm out of brackets, you move down the hierarchy: a power becomes multiplication; a root becomes division; a multiplication becomes addition; and a division becomes subtraction. (If you're trying to turn something *into* a single logarithm, you move upwards – towards the brackets.)

Here are a couple of other facts you need to know, for any positive value of a that isn't 1:

- $\log_a(a) = 1$
- $\log_a(1) = 0$

The first of those is because $a = a^1$, and the second is because $1 = a^0$.

The last log thing you need to know: you're only allowed to put positive numbers into a logarithm; $\log_a(0)$ is undefined (ask your calculator: you'll get a 'math error', whatever a you pick). The same goes for $\log_a(-5)$ or any other negative number.

Expressing logarithms in a simpler form

In Core 2, you're often asked to write down the value of a logarithm (such as $\log_4(64)$) or to turn a mess like $2\log_a(5) - 3\log_a(2) + \log_a(3)$ into a single logarithm in terms of a.

To find the value of a logarithm like the first one, you *could* type it into your calculator, write down 3 and move on. In fact, that's exactly what I suggest you do if they ask you anything so simple in the exam. However, while you're studying, you'll do better to figure out exactly what's going on and ask yourself how you'd get there if your calculator battery ran out.

You might start by saying that $\log_4(64)$ means 'What value of x solves $4^x = 64$?' If you don't spot that's 3, you need to memorize your powers; that's something that might come up in Core 1. The drudgery way of doing this problem is to work out powers of 4: $4^2 = 16$; $4^3 = 64$, so $x = 3$. There are alternative methods – for instance, you could factorise 64 and make 4s out of the factors.

The second example is trickier, and I'll talk you through it.

1. **Deal with any multiplications or divisions by bringing the multiplier into the bracket.**

   From $2\log_a(5) - 3\log_a(2) + \log_a(3)$, you get

   $$\log_a(5^2) - \log_a(2^3) + \log_a(3)$$

2. **Deal with any additions or subtractions by turning them into multiplications and divisions.**

   You now have $\log_a\left(\dfrac{5^2 \times 3}{2^3}\right)$.

3. **Work that out.**

   It's $\log_a\left(\dfrac{75}{8}\right)$, which you should leave as a fraction unless you're told otherwise.

Solving simple log equations

When you have an equation like $\log_a(x) + \log_a(3) = 2\log_a(6x)$ to solve, the trick is to make each side into a single logarithm, with everything inside the brackets. Here's how:

1. **First, deal with any multiplication or division.**

   On the right, you have a $2\log_a(6x)$, which you can quickly turn into $\log_a(36x^2)$.

2. **Deal with any addition or subtraction.**

   On the left, $\log_a(x) + \log_a(3) = \log_a(3x)$.

3. **Both sides should now be single logarithms, so do base-to-the-power on both sides.**

   This turns $\log_a(3x) = \log_a(36x^2)$ into $3x = 36x^2$.

4. **Solve this!**

   $$0 = 36x^2 - 3x$$
   $$0 = 3x(12x - 1)$$
   $$x = 0 \text{ or } x = \frac{1}{12}$$

5. **Check your answers make sense.**

   The answer $x = 0$ is invalid because $\log_a(0)$ is undefined; $x = \frac{1}{12}$ is the only answer.

Logs and numbers together

When you've got a whole batch of logarithms equal to each other, it's the work of minutes to combine them, get rid of the logs and solve whatever you can salvage from the rubble.

The trouble comes when you have numbers thrown into the mix. There are two pretty good methods for dealing with them – I have a slight preference for the everything's-a-logarithm method because you do your rearrangement up front and end up with something fairly nice, but really, whichever you're comfortable with is A-OK.

The numbers-on-one-side method

Putting the numbers on one side and everything else on the other is the traditional method for tackling the sort of logarithm problem that asks you to solve something like $\log_3(9x - 2) - 2 = 2\log_3(x)$. Here's how it goes:

1. **Rearrange so all the logs are on one side and all the loose numbers are on the other.**

$$\log_3(9x-2)-2\log_3(x)=2$$

2. **Combine the logs into a single logarithm.**

$$\log_3\left(\frac{9x-2}{x^2}\right)=2$$

3. **Select the right base, and raise this to the power of each side.**

Here, the base is 3, and raising it to the power of each side gives you

$$\frac{9x-2}{x^2}=9$$

4. **Solve the resulting equation.**

Set the equation equal to 0 and factorise:

$$9x-2=9x^2$$
$$9x^2-9x+2=0$$
$$(3x-2)(3x-1)=0$$
$$x=\frac{2}{3}\ \text{ or }\ x=\frac{1}{3}$$

5. **Check your answers are valid.**

Neither of them breaks the original equation, so you're good.

The everything's-a-logarithm method

The everything's-a-logarithm question hinges on knowing that $\log_a(a)=1$, which means you can always multiply a number by $\log_a(a)$ without changing its value. Tackling the same example as before, $\log_3(9x-2)-2=2\log_3(x)$, here's how this method goes:

1. **Multiply any loose numbers by $\log_a(a)$, where a is the base you're working with.**

Here, $a=3$, so $\log_3(9x-2)-2=2\log_3(x)$ becomes

$$\log_3(9x-2)-2\log_3(3)=2\log_3(x)$$

2. **Rearrange so everything is positive, because who needs minus signs?**

$$\log_3(9x-2)=2\log_3(x)+2\log_3(3)$$

3. Combine the logs on each side, where necessary.

$$\log_3(9x-2) = \log_3(9x^2)$$

4. Raise a to the power of each side.

$$9x-2 = 9x^2$$

5. Solve the resulting equation and check your answers.

Rearrange to get the same quadratic as before and solve.

(*Note:* Step 2 is completely optional, by the way. It just makes things a bit tidier.)

Logarithmic simultaneous equations

Simultaneous equations involving logarithms are a) rare but b) fair game. They're more fiddly than they are difficult, though! Usually, the best way to solve them is to get one of the variables on its own, substitute into the other equation and then solve. Sometimes, you can be a bit smarter – for instance, if you have $\ln(x)$ in both equations, you may be able to use *that* as your variable – but that's more of a once-in-a-while time-saver than a general rule.

A Core 2 example might ask you to solve $\log_5(x) + \log_5(y) = 2$ and $\log_5(6x-y) = 1$ simultaneously. Here's how I'd do it:

1. Get rid of the logs in the first equation.

$$\log_5(x) + \log_5(y) = 2$$
$$\log_5(xy) = 2$$
$$xy = 5^2$$
$$xy = 25$$

2. Get rid of the logs in the second equation.

$$\log_5(6x-y) = 1$$
$$6x-y = 5^1$$
$$6x-y = 5$$

3. Solve the new simultaneous equations.

$y = 6x - 5$ from the second equation, so $x(6x-5) = 25$ from the first. $6x^2 - 5x - 25 = 0$, which factorises as $(2x-5)(3x+5) = 0$, so $x = \frac{5}{2}$ or $x = -\frac{5}{3}$, which is invalid. If $x = \frac{5}{2}$, $y = 10$.

Changing bases

If the examiners are in an especially bad mood, they might ask you something involving base trickery. (Almost all logs questions you're likely to see involve a single, specified base – or at worst, a variable base that isn't especially relevant.) What's base trickery? It's giving you logs with different or unknown bases.

For instance, you might be told $\log_2(x) = \log_4(x+2)$ and be asked to find x. Nightmare! Luckily, there's a change-of-base formula:

$$\log_a(b) \equiv \frac{\log_c(b)}{\log_c(a)}$$

For this example, $\log_4(x+2)$ can be rewritten as $\dfrac{\log_2(x+2)}{\log_2(4)}$, and the bottom simplifies to 2. The equation is now $\log_2(x) = \dfrac{\log_2(x+2)}{2}$. Double both sides (to tidy up the fraction); you get $2\log_2(x) = \log_2(x+2)$. Then raise 2 to the power of both sides to get $x^2 = x+2$, or $x^2 - x - 2 = 0$, which factorises as $(x+1)(x-2) = 0$, so $x = -1$ (which is invalid) or $x = 2$, which is the only answer.

Another type of base trickery question might ask you to find x in $\log_x(9) = \log_3(x)$. That's trickier, but you can use the same rule to turn $\log_x(9)$ into $\dfrac{\log_3(9)}{\log_3(x)}$. The top of that is 2, which means you have $\dfrac{2}{\log_3(x)} = \log_3(x)$, or $2 = \left(\log_3(x)\right)^2$. You can solve that: $\log_3(x) = \pm\sqrt{2}$, so $x = 3^{\sqrt{2}}$ or $x = 3^{-\sqrt{2}}$, both of which are valid answers.

Solving tricky logs questions

To finish this section, I take you through some of the trickier logarithm questions the examiners may throw at you. Here's the general process for solving them:

1. **Combine the logarithms on each side into one bracket as quickly as possible.**

2. **Get rid of the logs carefully.**

3. **Solve what's left over.**

4. **Make sure your answers make sense.**

Solving for several separate logs

An exam-style logs question might ask something like this:

Given $-2 < x < 2$ and $\log_6(x+2) - 2\log_6(2-x) = 1$, find the value of x to three significant figures.

Here's what to do:

1. **Turn the logs on the left into a single logarithm.**

   First, the one with a 2 in front of it becomes $\log_6\left((2-x)^2\right)$; then the minus becomes a divide: $\log_6\left(\dfrac{x+2}{(2-x)^2}\right) = 1$.

2. **Now raise both sides to the power of the base to get rid of the log.**

   $$\frac{x+2}{(2-x)^2} = 6^1$$

   $$\frac{x+2}{(2-x)^2} = 6$$

3. **Now rearrange and solve.**

   $$x+2 = 6(2-x)^2$$
   $$x+2 = 6\left(x^2 - 4x + 4\right)$$
   $$x+2 = 6x^2 - 24x + 24$$
   $$0 = 6x^2 - 25x + 22$$

   I'm going to take a punt and say that doesn't factorise, so I'll throw the quadratic formula at it: $x = \dfrac{25 \pm \sqrt{625-528}}{12} = \dfrac{25 \pm \sqrt{97}}{12}$. That gives you $x = 1.26$ or $x = 2.90$, to three significant figures.

4. **Check your answers are valid.**

   Here, 2.90 is outside the domain you were given (you were told $-2 < x < 2$; also, $2 - 2.90$ is negative, so its logarithm is undefined). The only valid answer is 1.26.

Letters up top

The last kind of logarithm question you can expect is the 'letters up top', in which you have to solve for an unknown power. You know how to do something like $5^x = 12$ (you just take logs base 5 and get $x = \log_5(12) \approx 1.544$), but what about something like $5^x = 2 \times 7^{x-2}$? It's actually not *all* that dissimilar.

1. **Take logarithms (whatever base makes you happy).**

   If 10 makes you happy,

   $$\log_{10}\left(5^x\right) = \log_{10}\left(2 \times 7^{x-2}\right)$$

2. **Get as much as you can out of the brackets.**

   $$x \log_{10}\left(5\right) = \log_{10}\left(2\right) + \left(x-2\right)\log_{10}\left(7\right)$$

3. **Rearrange to get your xs on one side and everything else on the other.**

   $$x \log_{10}\left(5\right) - x\log_{10}\left(7\right) = \log_{10}\left(2\right) - 2\log_{10}\left(7\right)$$

4. **Factorise and divide.**

   The left-hand side is $x\left(\log_{10}\left(5\right) - \log_{10}\left(7\right)\right)$, so

   $$x = \frac{\log_{10}\left(2\right) - 2\log_{10}\left(7\right)}{\log_{10}\left(5\right) - \log_{10}\left(7\right)}$$

If they want an exact answer, this will do. If they want a decimal approximation, stick it in the calculator: I get 9.5065 or so.

You'll end up with the same decimal answer whichever base you pick for your logarithms!

Making Sense of Euler's constant, e

I've seen some of my MathsJam buddies wearing a t-shirt that reads, 'e: for when π isn't geeky enough'. The e stands for Euler – Leonhard Euler (it rhymes with 'boiler'), probably the most prolific mathematician ever. One of the rules for naming things in maths is that you name it after the *second* person to discover it, or else almost everything would be named after Euler. It's ironic, then, that he didn't discover e (depending on who you believe, it was William Oughtred, John Napier or Gottfried Leibniz). He did, however, do a lot of work on it, much of which you'll see either in the first year of university maths or in the Further Pure modules.

The number e is involved in Euler's identity (probably discovered by Roger Cotes or Johann Bernoulli):

$$e^{\pi i} + 1 = 0$$

where π is pi and i is the imaginary square root of –1; this identity is usually voted as the most mathematically beautiful equation. (Personally? I prefer $D = S + X - N_C - 1$, but I may be biased.)

So what is e? And why is it so important? Read on.

Understanding that e is just a number (but a special one)

There are several ways of defining the constant e, which has a value of about 2.718281828459045. (Yes, I do know it, off the top of my head, correct to 16 significant figures. Why do you ask?) The best approximation is this:

$$e = \frac{1}{0!} + \frac{1}{1!} + \frac{1}{2!} + \frac{1}{3!} + \frac{1}{4!} + \ldots = 1 + 1 + \frac{1}{2} + \frac{1}{6} + \frac{1}{24} + \ldots$$

which gives you the number above. (Incidentally, the ! sign means *factorial*: the product of all the natural numbers up to and including that number – so $4! = 4 \times 3 \times 2 \times 1$. And $0! = 1$ – just roll with it.)

This number has the special property that if you differentiate e^x (which you'll do in Chapter 14), you get e^x. It's the only function (apart from 0) that differentiates to itself. For that reason, it crops up all the time in calculus – and it's the (ahem) natural choice for a logarithmic base, so much so that logarithms with base e are called *natural logs* and are denoted by $\ln(x)$.

Converting between powers

It's rare to be asked explicitly to convert expressions between one power and another, but you still need to be able to do it. If you have something like 10^{3x} to differentiate or integrate (Chapters 14 and 16, respectively), you need to turn that value into a power of e. If you have different power bases in an equation, you probably want to turn them into the same base. And if you had to pick a base? That's right, you'd pick e.

The main fact you need to know for power conversion is one you know already: $a = e^{\ln(a)}$, because e-to-the-power-of and the natural logarithm are inverse functions – they undo each other. So if you wanted to convert 10^{3x} into a power of e, you'd do the following:

1. **Write the base as $e^{\ln(a)}$.**

   In this case, you write the base, 10, as $e^{\ln(10)}$, so

   $$10^{3x} = \left(e^{\ln(10)}\right)^{3x}$$

2. **Apply power laws: when you raise one power to another, you multiply the powers.**

   $$10^{3x} = e^{3x\ln(10)}$$

It's now in a form you can do something useful with!

If you had different powers on each side of an equation, you could use this trick to make them both powers of e. (If *all* you have is two different powers on each side of the equation, you should just take logs, but something more complicated could be a disguised quadratic – which you see more of later in this chapter!)

For example, if you need to solve $4^x e^{6x} = 5^7$, you could approach it like this:

1. **Take logs of both sides and simplify.**

   $\ln\left(4^x e^{6x}\right) = \ln\left(5^7\right)$, so $x\ln\left(4\right) + 6x = 7\ln\left(5\right)$.

2. **Get all the *x*s together and rearrange.**

   $x\left(\ln\left(4\right) + 6\right) = 7\ln\left(5\right)$, so $x = \dfrac{7\ln\left(5\right)}{\ln\left(4\right) + 6}$.

3. **Depending on what you're asked for, you can leave your answer like that or give it to as many decimal places as you're asked for.**

   I get 1.525 to three decimal places. Putting this x-value back into the original equation (using your *Ans* button rather than the rounded decimal) as a check gives you 78,125 on the left-hand side, which is 5^7 – so I'm happy that 1.525 is the correct answer.

Solving things with e in them

Repeat after me: it's just a number! In fact, if you're going to raise any number to an unknown power, e is by far the easiest to deal with (except, possibly, 0 or 1). It's comparatively easy to differentiate, you have a natural log button on your calculator to get rid of it if you need to, you know your power laws... using e is just a case of piecing everything together.

The number e crops up a lot in real-life models – population growth or decline, cooling and heating bodies, radioactive decay, medication uptake and so on. I've even seen plagues of locusts modelled with e (but that was on a Solomon paper). You're almost certain to see something e-related in Core 3, and you need to be able to deal with it.

Typical models

A standard exponential model question might give you a nice hot cup of coffee, the temperature of which (in degrees Celsius) is modelled as $T = 20 + Ae^{-kt}$, where A and k are constants and t is measured in minutes. You may be told the initial temperature (let's say 85°C) and the temperature at some other time (say, after 10 minutes, it's at 40°C) and be asked to find A and k. Here's what you'd do:

1. **Use the initial value (when $t = 0$) to work out A.**

   $85 = 20 + Ae^0$, so $A = 65$.

2. **Use this and the second piece of information to work out k.**

   $40 = 20 + 65e^{-10k}$, which you can rearrange to get $\frac{20}{65} = e^{-10k}$. Taking logs, $\ln\left(\frac{20}{65}\right) = -10k$, so $k = \frac{1}{10}\ln\left(\frac{13}{4}\right)$ if you mess around with the minus sign.

$-\ln(x) \equiv \ln\left(\frac{1}{x}\right)$, using your power laws.

Variations on the theme include the following:

- ✔ **Finding the temperature at some other time:** All you need to do is use the values you've worked out and let t be the time you're given.

- ✔ **Finding how long it takes to reach some other temperature:** Let T be your target temperature and rearrange to find t.

- ✔ **Finding the rate of change of temperature at some time:** Differentiate and put your value of t where it belongs so you can find $\frac{dT}{dt}$.

- ✔ **Finding when the rate of change is a particular value:** You might be getting the hang of this by now! Differentiate, set $\frac{dT}{dt}$ to the given value, rearrange and solve for t.

I've used 'temperature' all the way through here, although your model could just as well refer to a price, a population, an electric current, a concentration of medication in a bloodstream, or any number of other things – as I said, exponential models have an extremely wide range of applications.

Half-lives and similar

Because exponential functions grow at a variable (but predictable) rate, examiners often think it's a good laugh to ask students when the value of the function has doubled or halved or reached some other arbitrary proportion of its original value. For example, you might have a function describing the amount, M, of radioactive material you have lying about in your shed: $M = M_0 e^{-0.2t}$, where M_0 is the amount you had originally and t is the number of years it's been lying there. (**Warning:** Don't try this at home.)

How long (an examiner might ask) will it be until the mass has decreased by 10 per cent? Here's how to answer:

1. **Work out how much of the original amount you need.**

   If you've lost 10 per cent of the original M_0, you now have $0.9M_0$ left.

2. Set up the equation and divide by M_0.

You start with $0.9M_0 = M_0 e^{-0.2t}$ and get $0.9 = e^{-0.2t}$.

3. Take natural logs.

$$\ln(0.9) = -0.2t$$

4. Solve for t.

$$t = -5\ln(0.9)$$

5. If necessary, convert this into seconds, minutes, years or whatever they ask for.

This would be around 0.527 years.

To convert a decimal number of hours into hours, minutes and seconds, you can use the 'button with commas on' on your calculator. It's really for converting degrees with a decimal part into degrees, minutes and seconds, but it works just as well for hours.

Long-term behaviour

A common twist in a Core 3 or Core 4 question involving exponentials is to ask what happens to a variable in the long term – or to ask you to show that a variable never reaches a given value. All this kind of question is asking is 'Do you know that when x gets large, e^{-x} gets really, really small?' (For example, e^{-20} is about $\dfrac{1}{500,000,000}$, which is a pretty small number in my book. And guess what! This *is* my book.)

With that in mind, here's how you'd find the long-term behaviour of $y = \dfrac{400e^t}{5 + 10e^t}$:

1. Make sure you have only negative powers of e.

Here, divide top and bottom by e^t to get $y = \dfrac{400}{5e^{-t} + 10}$.

2. Replace all of the e^{-t}'s with 0.

That gives you $y \to \dfrac{400}{10}$ as $t \to \infty$.

3. Simplify.

As t gets large, y gets close to 40.

In Step 2, using an equals sign would be technically incorrect; t never reaches infinity, and y never reaches $\dfrac{400}{10}$, so 'equals' is inappropriate. Instead, you should really use an arrow to say 'approaches' or 'tends to'. You're unlikely to lose marks for using an equals sign at this stage, but you should develop good habits now!

If you want to show that a function with *es* in it never gets above or below a given value, you can approach it the same way – just figure out what the long-term behaviour is and state that the function is always above (or below, depending on the question) that value.

An alternative way to show a certain value is impossible is to try to solve for the given value – you'll either get no solution (you'll need to take the logarithm of a negative number) or a nonsense answer (for example, a negative time or an *x* outside of the given domain.)

This technique also crops up in function questions: you're frequently asked to give the range of a function involving an *e*. If it's a simple function (such as $f(x) = 2 + 3e^x$), you just need to think about the biggest and smallest $f(x)$ can get: if *x* equals a negative number of enormous magnitude (say, –1,000), you get something a tiny bit more than 2; for a large value of *x*, $f(x)$ gets enormous. The range here would be $f(x) > 2$. Chapter 9 gives a more detailed treatment of domains and ranges.

Watching out for booby-traps

Although most questions with an *e* in are straightforward, once in a while, the examiners will throw in a wrong 'un just to keep you on your toes. If you *are* on your toes, you'll deal swiftly with simultaneous equations, ruthlessly remove the disguise of quadratic equations, and masterfully merge multiple *es*.

In short, anything you can do with 'normal' equations, you want to be able to do with *es*. With *ease*. Get it? Oh, never mind.

Simultaneous equations

It's unusual to get a straight-up 'solve these simultaneous equations' question in Core 3, but you quite often need to find the intersection of two curves, either of which may have an *e*. The method I recommend is just the same as for any other simultaneous equations: isolate something you can eliminate from both equations, and solve what's left over.

For example, to work out where $y = \frac{1}{2}\left(e^x - e^{-x}\right)$ and $y = \frac{1}{4}e^x$ intersect, here's what you'd do:

1. **Notice that the left-hand sides are equal, so the right-hand sides are also equal.**

   That means $\frac{1}{2}\left(e^x - e^{-x}\right) = \frac{1}{4}e^x$.

2. Solve this.

$2e^x - 2e^{-x} = e^x$, so $e^x = 2e^{-x}$.

Multiplying by e^x gives you $e^{2x} = 2$, so $2x = \ln(2)$ and $x = \frac{1}{2}\ln(2)$.

3. Go back to the original equations to find *y*.

$y = \frac{1}{4}e^x$, which is $\frac{1}{4}\sqrt{2}$, because $\frac{1}{2}\ln(2) = \ln\left(\sqrt{2}\right)$ and $e^{\ln(\sqrt{2})} = \sqrt{2}$.

The curves intersect at $\left(\frac{1}{2}\ln(2), \frac{\sqrt{2}}{4}\right)$.

Depending on what you're trying to solve, you may need to rearrange the equations to get them into a 'nice' form.

Disguised quadratics

For some reason, disguised quadratics with an *e* in seem much harder to spot than quadratics in other forms. For some students, it stems from not being 100 per cent comfortable with the power laws, so I recap the important ones here:

✔ $a^{bc} \equiv \left(a^b\right)^c$, and in particular, $e^{2x} = \left(e^x\right)^2$

✔ $a^{-b} \equiv \frac{1}{a^b}$, and in particular, $e^{-x} = \frac{1}{e^x}$

Knowing these laws gives you some clues about how to approach anything awkward involving *e*s: try substituting $y = e^x$ (or something equally sensible) and see what comes out of the mess. For example, with something like $\frac{e^x + 7}{e^{-x} + 4} = 2$, here's what I'd do:

1. Rearrange so it's in a nicer form.

$$\frac{e^x + 7}{e^{-x} + 4} = 2$$
$$e^x + 7 = 2\left(e^{-x} + 4\right)$$
$$e^x = 2e^{-x} + 1$$

2. Substitute $y = e^x$.

$$y = \frac{2}{y} + 1$$

3. Solve this using the usual techniques.

Multiply both sides by *y* to get $y^2 = 2 + y$; then rearrange so that $y^2 - y - 2 = 0$. This factorises as $(y-2)(y+1) = 0$. Either $y = 2$ or $y = -1$.

4. Solve for the original.

$y = e^x$, so either $e^x = 2$ or $e^x = -1$, which is impossible. The only solution is $x = \ln(2)$.

5. Check with the original equation.

$$\frac{e^{\ln(2)} + 7}{e^{-\ln(2)} + 4} = 2$$

On the left-hand side, the top is $2 + 7$, and the bottom is $\frac{1}{2} + 4$; the fraction is $\dfrac{9}{\left(\dfrac{9}{2}\right)}$ or 2, which matches the right-hand side, so you're happy with $x = \ln(2)$ as an answer.

More than one e

Dealing with more than one e in a single equation is something I'm forever seeing good A level students struggle with, so I cover it here even though there are no new skills required. The kind of question I mean is when you need to solve, say, $100 = \dfrac{2{,}200e^{5t}}{9 + 16e^{5t}}$. Here are the steps I'd take:

1. Ask, 'what's ugly?'

Here, it's the fraction, so multiply both sides by the bottom to get

$$100\left(9 + 16e^{5t}\right) = 2{,}200e^{5t}$$

2. Get your unknowns on one side and everything else on the other.

$$900 + 1{,}600e^{5t} = 2{,}200e^{5t}$$
$$900 = 600e^{5t}$$

3. From here, it ought to be routine to rearrange and solve.

$\dfrac{3}{2} = e^{5t}$, so $5t = \ln\left(\dfrac{3}{2}\right)$ and $t = \dfrac{1}{5}\ln\left(\dfrac{3}{2}\right)$.

Ninety-nine times out of a hundred, a solution like that is exactly what the examiners want – but if they ask for a certain number of decimal places, you should obviously give them what they want (0.081 or so).

You can make the algebra a bit more familiar if you like by replacing e^{5t} with, say, y and solving for that – but remember to solve for t at the end!

Chapter 6

Playing with Polynomials

*T*he *polynomial* is – arguably – one of the most important kinds of functions in maths. (I say 'arguably', but please don't argue it with me.) So what is a polynomial? Here are some examples: $p(x) = 0$ is a polynomial. So are $q(x) = k + x$ (where k is a constant), $r(x) = x^2 + 3x - 2$, $s(y) = y^9 + 3y^4 - \frac{9}{4}y + \pi$ and $t(x) = 1 - \frac{1}{2}x^2 + \frac{1}{24}x^4 - \frac{1}{720}x^6$. Here's what they have in common:

   ✔ They're functions of one variable – usually x or y, but that's just convention; there's no reason it can't be θ or z or k, for all I care.

   ✔ Every term in the polynomial is of the form ax^n, where a is a constant and n is a nonnegative integer – there are no negative or fractional powers.

   ✔ The terms can be written in any order (with the highest power first, last or somewhere in the middle). Conventionally, the terms are kept in the order of their powers, but sometimes the powers start small and get bigger, and sometimes they go the other way around.

   ✔ There are no other functions of x involved – no cosines, no exponentials, no xs in the power or on the bottom of fractions. In short, no funny business.

The *degree* of a polynomial is the highest power of x involved – so $t(x)$ is a polynomial of degree 6, and $s(y)$ has degree 9. Unless you're rather unlucky or do Further Maths, these are the last polynomials of degree 6 or 9 you're likely to see. A level deals mainly with polynomials of degree 3 or smaller.

Polynomials of degree 3 are called *cubics*; degree 2 is a *quadratic*; degree 1 is a *linear* expression; and degree 0 – like $p(x)$ in the first paragraph – is a *constant*.

You do see higher-order polynomials at A level – especially in binomial expansion (which is in Chapter 8), but the chances that you'll have to solve a polynomial with a degree of more than 4 are very slim indeed.

Completing the Square

Completing the square is a technique for working with quadratics that, in honesty, I've never had much use for. It's one of those it's-in-the-test-so-you-need-to-know-it topics rather than something that's going to seriously hinder your mathematical progress if you don't know it.

It has two uses at A level: you can use it to solve a quadratic equation (although usually, other methods are less error-prone and time-consuming) or to find the turning point (and again, there are other ways to do that). It's also the source of the quadratic formula, which is one argument I can see for keeping it from a usefulness point of view. It also has applications in integration, but those are most likely *way* off in the future for you.

For all that I like to grumble about how much I'd like to see it kicked off the syllabus, the mathematician in me does appreciate its elegance!

Following the basic method

There are several methods for completing the square, so the chances of your textbook's version matching with mine are slim. Here's how I recommend putting a quadratic such as $3x^2 + 8x - 7$ into the form $p(x-q)^2 + r$.

1. **Start by expanding the template.**

   You get $px^2 - 2pqx + pq^2 + r$.

2. **Match coefficients of x^2.**

   Your quadratic has 3 of them, and your template has p of them, so $p = 3$.

3. **Match coefficients of x.**

   Your quadratic has 8 of them, your template has $-2pq$ of them, and $p = 3$, so $q = -\frac{4}{3}$.

4. Match coefficients of the units.

Your quadratic has -7 of them, and your template has $pq^2 + r$ of them; $pq^2 = 3 \times \left(-\frac{4}{3}\right)^2 = \frac{16}{3}$, so $r = -\frac{37}{3}$.

5. Write down your answer.

$$3x^2 + 8x - 7 \equiv 3\left(x + \frac{4}{3}\right)^2 - \frac{37}{3}$$

Solving a quadratic by completing the square

Once you have a quadratic in completed-square form, you can find the values of x that give 0 without too much extra work. For example, if you arrange the quadratic $2x^2 - 12x + 3 = 0$ as $2(x-3)^2 - 15 = 0$, you can solve for x in a fairly instinctive way:

1. Rearrange to get the squared bracket on its own.

$$(x-3)^2 = \frac{15}{2}$$

2. Find the square roots of both sides.

$$x - 3 = \pm\sqrt{\frac{15}{2}}$$

3. Rearrange to get x on its own.

$$x = 3 \pm \sqrt{\frac{15}{2}}$$

If the thing on the right in Step 1 is negative, then there are no solutions (it's related closely to the discriminant, in fact); if it works out to 0, there is a repeated root.

Otherwise, you're looking for two solutions, and if you find only one, something has gone wrong. I'll tell you what's gone wrong: you missed out the $\pm$ in Step 2. Go back and put it in.

Finding the vertex

Finding the *vertex* – the lowest or highest point on a quadratic curve, or its turning point – is simple: if you've managed to rearrange the quadratic into the form $p(x-q)^2 +r$, the vertex is at (q,r).

There are a couple of ways to see why that is. The first is a graphical argument, using some of the ideas from Chapter 10. If you start with the graph $y = x^2$, with its vertex at $(0,0)$, and move the graph q units to the right, it becomes $y = (x-q)^2$, with its vertex at $(q,0)$. Stretching it vertically by a scale factor of p turns the graph into $y = p(x-q)^2$ but doesn't move the vertex. Lastly, moving the graph up by r turns the equation into $y = p(x-q)^2 +r$ and moves the vertex to (q,r).

You can also make an algebraic argument for why the minimum value of the quadratic occurs when $x = q$: the smallest the squared bracket ever gets is 0 (because it's squared, it can never give a negative answer), and that happens when $x = q$. Putting $x = q$ into the whole equation gives you $y = r$, which means the vertex is at (q,r).

Understanding where the quadratic equation comes from

Suppose you have to solve the equation $ax^2 + bx + c = 0$, and the quadratic formula, for some reason, has been lost to humankind. How do you regenerate it? Humankind is relying on you!

Well, you start by completing the square, which is a bit awkward, but go for it. Start with the template, $p(x-q)^2 +r$, and expand to get $px^2 - 2pqx + pq^2 +r$, which needs to match up with $ax^2 + bx + c$. That means $p = a$, $-2pq = b$ and $pq^2 +r = c$. Solving those gives you $q = -\dfrac{b}{2a}$ and $r = c - \dfrac{b^2}{4a}$.

Your quadratic is now

$$a\left(x+\frac{b}{2a}\right)^2 +c - \frac{b^2}{4a} = 0$$

Move the constants to the other side and divide by a:

$$\left(x+\frac{b}{2a}\right)^2 = \frac{b^2}{4a^2} - \frac{c}{a}$$

The fractions on the right can be combined as $\dfrac{b^2 - 4ac}{4a^2}$ – that top should look familiar! Take a square root on both sides:

$$\left(x + \frac{b}{2a}\right) = \pm\sqrt{\frac{b^2 - 4ac}{4a^2}}$$

The clever bit: you know from Chapter 3 that you can split roots over a fraction, so $\sqrt{\dfrac{b^2 - 4ac}{4a^2}} = \dfrac{\sqrt{b^2 - 4ac}}{2a}$. Take away $\dfrac{b}{2a}$ from each side of the equation, and you have

$$x = -\frac{b}{2a} \pm \frac{\sqrt{b^2 - 4ac}}{2a}$$

which – when you turn it into a single fraction – gives you the quadratic formula.

Thank you for saving the world.

Factorising and Solving Simple Polynomials

If you need to factorise a quadratic equation, you're in the wrong place – Chapter 3 is where you need to be! If you have a complicated cubic (or higher), then Chapter 7 is your friend. Right here is the room for cubics like $x^3 + 4x^2 - 5x$ – ones that don't have a number at the end.

This kind of cubic crops up in Core 1 all the time, and it isn't too tough to deal with.

You're also in the right place if you've got an equation wearing a silly moustache and a funny nose: disguised quadratics also appear from Core 1 onwards, and you need to know how to get their disguise off, solve them and get their disguise back on as if nothing had happened.

Finding simple factors

Given a cubic expression like $x^3 + 4x^2 - 5x$, your first instinct should be to ask whether you can factorise it. If there's no constant term on the end, then you definitely can: there's a common factor of x in every term, so you can rewrite it as $x\left(x^2 + 4x - 5\right)$.

Your next instinct should be to ask whether that quadratic bracket factorises. Indeed it does: it's $(x+5)(x-1)$, so the whole expression is equivalent to $x(x+5)(x-1)$.

Why is that useful? Because factorised expressions are happy expressions. This happiness manifests itself in several ways:

- ✔ **It's easy to find the values of x that make the expression 0.** You just look for where any of the factors is 0: $x(x+5)(x-1)$ has solutions at $x=0$, $x=-5$ and $x=1$.

- ✔ **It's easy to sketch the associated graph once you know where the polynomial is 0.** For example, the graph of $y = x(x+5)(x-1)$ crosses the x-axis at $(0,0)$, $(-5,0)$ and $(1,0)$.

- ✔ **It's easy to evaluate the expression.** Working out three simple sums with a single x in each and multiplying the answers is less involved than calculating several powers and adding them up. For example, if $x=\frac{1}{2}$, then the factorised version gives you $\left(\frac{1}{2}\right)\left(\frac{11}{2}\right)\left(-\frac{1}{2}\right) = -\frac{11}{8}$. The original version gives you $-\frac{11}{8}$ too, but it requires a lot more work: $\left(\frac{1}{2}\right)^3 + 4\left(\frac{1}{2}\right)^2 - 5\left(\frac{1}{2}\right) = \frac{1}{8} + \frac{4}{4} - \frac{5}{2} = \frac{1}{8} + \frac{8}{8} - \frac{20}{8} = -\frac{11}{8}$. Multiplying fractions beats adding them, any day!

Taking off a quadratic's disguise

A *disguised quadratic* is something you can turn into a quadratic equation by careful rearrangement and clever variable-swapping. Quadratics really are masters of disguise – here are some of the ways they can sneak up on you. This list isn't exhaustive, but it gives you an idea of what to look for:

- ✔ **Polynomials where one power is double the other:** For instance, if $x^6 + 9x^3 - 10$, you could replace x^3 with y, because $x^6 = \left(x^3\right)^2 = y^2$, leaving you with a quadratic.

- ✔ **Expressions with the same power on the top and the bottom:** You could rearrange $y^2 + 7 = \frac{6}{y^2}$ into a disguised quadratic by multiplying everything through by y^2, which you'd then replace with z.

- ✔ **Mixed trig functions:** If you see something with $\sin^2(\theta)$ and $\cos(\theta)$ (or similar) in it, you can bet your last protractor you're going to end up using identities to get rid of the sine and end up with a disguised quadratic. See Chapter 12 for more.

✔ **Nasty exponentials:** The first two bullets are just as true for e^x as they are for x^3 – if you've got one power that's double another (such as $e^{4x} - 5e^{2x} + 4$, in which case you'd set $y = e^{2x}$), or the same power on the top in one term and on the bottom (or as a negative power, which amounts to the same thing) in another (for example, $e^{3x} - 4 + 4e^{-3x}$, where you'd use $y = e^{3x}$), be on the lookout for a quadratic.

✔ **Very nasty powers:** Something like $9^x - 3^{x+1} + 2$ looks impossible – until you realise you can use power laws to work it out (replacing 3^x with y). Read on for details.

It doesn't matter which letter you use to replace the nastiness – conventionally, you'd pick something you've not used yet, from towards the end of the alphabet, but you'd not get marked down for using d or γ or anything else.

Here's how you'd tackle that last example, $9^x - 3^{x+1} + 2 = 0$:

1. **Get everything into the same base (in this case, 3).**

   $9^x = 3^{2x}$, because you can write it as $\left(3^2\right)^x$, and you can also rewrite the second term as a multiple of a power of 3: $3^{x+1} = 3 \times 3^x$, so the equation is

   $$3^{2x} - 3 \times 3^x + 2 = 0$$

2. **Now you have a double-power situation.**

   Make the substitution $y = 3^x$. You get

   $$y^2 - 3y + 2 = 0$$

3. **Solve the quadratic for y.**

   It factorises as $(y-1)(y-2) = 0$, so $y = 1$ or $y = 2$.

4. **Solve for x.**

   If $y = 3^x$ and $y = 1$, $x = 0$. The other solution, $y = 2$, gives you $3^x = 2$, so $x = \log_3(2) \approx 0.631$.

It's very easy to miss out Step 4. It's also easy to end up with wrong solutions (it's a good idea to make sure your final xs work in the original equation) or to miss possible solutions (especially with trigonometric questions). Be on your toes!

Counting Real Roots

The number of solutions a polynomial equation has is a really deep and beautiful part of mathematics – I'd list the Fundamental Theorem of Algebra as one of my favourite things – but, unfortunately, A level doesn't really look at that. It doesn't even go into complex numbers; if you ever think, 'There must be a piece missing from A level that would link several topics together,' you're right. It's complex numbers.

So you don't get to learn that every quadratic has two roots (although they may be the same, and they may be complex), every cubic has three, and every polynomial of degree *n* has *n* roots. Instead, you're told that a quadratic may have zero, one or two real roots. This is *true*, of course; it's just less tidy than the whole truth.

An expression in a single variable may have one or more *roots*: values that make the expression equal to 0. By contrast, an equation may have one or more *solutions*: values that make the equation true.

In Figure 6-1, there are three graphs: $y = x^2 + 1$, $y = x^2$ and $y = x^2 - 1$. The first has no real solutions for $y = 0$: the graph never crosses the *x*-axis. The second has one real solution, also known as a repeated root: the graph just touches the *x*-axis. And for the final graph, $x^2 - 1 = 0$ has two distinct real roots, because the graph crosses the *x*-axis twice. This section deals with how you tell the three cases apart, and it shows you how to work backwards from the number of solutions to work out the possible values of a constant.

Dealing with discriminants

If you have a quadratic expression written as $ax^2 + bx + c$, its *discriminant,* Δ, is $b^2 - 4ac$. Recognise it? It's the bit under the square root in the quadratic formula. The discriminant tells you how many solutions (or 'roots') there are to $ax^2 + bx + c = 0$:

- ✔ If $b^2 - 4ac > 0$, there are two distinct solutions.

- ✔ If $b^2 - 4ac = 0$, there is one solution (a repeated root).

- ✔ If $b^2 - 4ac < 0$, there are no real solutions.

If a quadratic equation has no real solutions, it has complex solutions – which, sadly for me but happily for you, you don't need to worry about.

This makes sense, if you think about the quadratic formula: if you try to take the square root of a negative number, your calculator complains that it's not possible, so the formula doesn't give you any answers. Meanwhile, the square root of 0 is 0, so if $\Delta = 0$, then you're adding and taking away 0 on the top – which means you get the same answer twice. Only when Δ is positive do you get two different solutions.

(a)

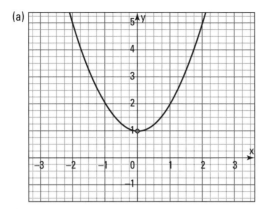

(b)

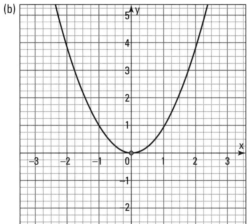

(c)

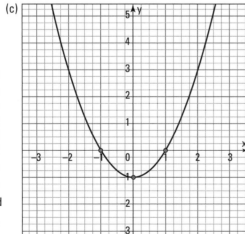

Figure 6-1:
Three qua-
dratics with
different
numbers of
real roots.

(a) $y = x^2 + 1$,

(b) $y = x^2$ and

(c) $y = x^2 - 1$.

Illustration by John Wiley & Sons, Inc.

Finding the numbers of solutions

Several types of questions involve the number of solutions to an equation (or system of equations), even when you exclude the ones that are simple applications of the discriminant.

To find out how many times a pair of simultaneous equations cross, you have two options: you may turn them into a single equation and find how many solutions it has, or you may sketch them both and see how many times they cross. If you go for Option B, be on the lookout for clues from earlier in the question about any points either of the curves goes through, because these will help your sketch no end.

Another type of question involving numbers of roots involves cubics (or, rarely, higher-order polynomials). Every (real) cubic equation has at least one real root and at most three. The challenge is to pick which of the three cases you have. You can approach this in many ways, but I recommend one of two, depending on whether your expression is factorised or not. In all cases, make sure you have an equation that's equal to 0 before you try to solve it.

Factorised

If an equation with a 0 on one side is factorised, every linear factor – an x or an $(x-k)$ – counts for one solution. A bracket raised to a power, like $(x-k)^3$, counts for a single solution, and a quadratic bracket you can't factorise further counts for as many real solutions as the quadratic itself has. Some examples:

- ✔ $(x-1)(x-2)(x-3) = 0$ has three solutions, one for each linear bracket.

- ✔ $x(x-2)^2 = 0$ has two solutions, one for the linear factor (x) and one for the squared bracket.

- ✔ $(x-2)(x^2+4) = 0$ has one solution – it comes from the linear factor; the quadratic factor has no real roots.

- ✔ Deep breath: $x(x-1)^2(x+2)^3(x-3)(x^2+4)(x^2-5) = 0$ has six solutions – one each from the linear factors, x and $(x-3)$; one each from the powered brackets, $(x-1)^2$ and $(x+2)^3$; two from the quadratic factor with two solutions, (x^2-5); and none from the quadratic factor with no solutions, (x^2+4). That's a polynomial of degree 11, by the way, far bigger than anything you'd expect to see in A level; it's for demonstration only!

Not factorised

If your expression that's equal to 0 *isn't* factorised, and assuming you can't factorise it, then you need to be a bit more cunning. A sketch of the relevant graph always helps, of course. A good strategy is to find and classify the turning points (if you see this with anything more complicated than a cubic, I'll. . . well, not eat my hat, because I need it to keep my head warm, but I'll drink some horrible tea). Once you've got your turning points, you can say the following:

- ✔ If there are no turning points or one turning point, there is only one real root to your cubic.

- ✔ Otherwise, if the *y*-value of both turning points is positive or negative, there is also only one real root.

- ✔ If the *y*-value of either turning point is 0, there are two roots. (One is a double root.)

- ✔ If the *y*-values of the two turning points have different signs, there are three roots.

Working with unknowns

A Core 1 question will quite frequently give you a quadratic equation with an unknown constant (usually *k*), tell you how many solutions the equation has and ask you what the possible values of *k* are. In this section, I show you how to deal with this kind of question when there's one real root (or a repeated root or equal roots, all of which mean the same thing). To deal with no real roots or two real roots, you need to know about inequalities, which are in the next section.

Suppose you're given the quadratic $kx^2 + 4x + (3k+1) = 0$ and told it has one real root. Here's how to find what *k* could be:

1. **Find the discriminant in terms of *k*.**

   Here, $b^2 - 4ac = 16 - 4k(3k+1)$, which is $-12k^2 - 4k + 16$.

2. **Solve for where the discriminant is 0.**

   There's a factor of –4 all the way through $-12k^2 - 4k + 16 = 0$, so you can simplify it to $3k^2 + k - 4 = 0$; this factorises as $(3k+4)(k-1) = 0$, so $k = 1$ or $k = -\frac{4}{3}$.

3. **If you were asked for the possible values of *k*, you'd be done. If you were asked for the possible real roots, you'd then need to solve the quadratic for each value of *k*.**

The original quadratic is $kx^2 + 4x + (3k+1) = 0$. If $k = 1$, then

$$x^2 + 4x + 4 = 0$$
$$(x+2)^2 = 0$$
$$x = -2$$

If $k = -\dfrac{4}{3}$, then

$$-\dfrac{4}{3}x^2 + 4x - 3 = 0$$

Multiplying everything by -3 gives you

$$4x^2 - 12x + 9 = 0$$
$$(2x-3)^2 = 0$$
$$x = \dfrac{3}{2}$$

If you have one real root, there will be two possible values for *k* (or, in *very* rare cases, one). If you have two real roots or none, then your answer will be one or two intervals – find out more in the following section!

Fighting Inequalities

Equations are statements of fact that are pretty easy to understand: the thing on one side of the '=' has the same value as the thing on the other side. $3x + 5 = 8$ means that the expressions '$3x + 5$' and '8' have the same value.

An *inequality* is also a statement of fact. Instead of saying, 'These two things have the same value', it allows expression to be greater. That's 'greater' in a mathematical sense of 'further to the right on the number line': -1 is greater than -2, and 7 is greater than -10.

You should know about four inequality symbols:

- $a > b$ means *a* is greater than *b*.
- $a \geq b$ means *a* is greater than or equal to *b* – or, in plain English, *a* is at least as big as *b*.

✔ $a < b$ means a is less than b.

  ✔ $a \leq b$ means a is less than or equal to b – or, in plain English, a is no bigger than b.

The small end of the inequality sign goes with the smaller number.

Inequalities work, in many respects, just like equations – but there are a few gotchas to look out for. In this section, I show you some of them.

Linear inequalities

A *linear* inequality is one where none of the variables are taken to a power or multiplied by each other. Or, put another way, variables are multiplied only by constants. Here are some examples of linear inequalities:

  ✔ $x > 0$

  ✔ $6y < 3$

  ✔ $3z + 4 \geq 9z - 2$

Solving an inequality usually means getting the variable on its own and related only to constants – the first example in the list is solved, but the other two aren't.

Algebraically, you can safely do anything with a linear inequality you can do with an equation *except* multiply or divide by a negative number. To solve the second example, you'd simply divide by 6 to get $y < \frac{1}{2}$.

I can't think of any reason you would *want* to square or square-root a linear inequality, but just in case you get the urge . . . squaring inequalities is dangerous, and you should avoid it if you possibly can. It can introduce wrong answers and mess with the direction of the inequality, so don't do it.

The third example is only slightly more involved: take away $3z$ from each side and add 2 to each side, and you have $6 \geq 6z$; then divide by 6 to get $1 \geq z$ (or $z \leq 1$). You need to be careful, though: if you had taken away $9z$ instead, you would have ended up with $-6z \geq -6$; dividing by -6 would give you something untrue ($z \geq 1$) – in fact, the $\geq$ would end up pointing the wrong way. That's a hint about how to deal with negatives in linear inequalities.

Be especially careful if you have an inequality that involves negatives. If you *must* multiply or divide by a negative number, be sure to flip the direction of the inequality sign.

Quadratic (and higher-order) inequalities

I think you might be able to guess what I mean by a *quadratic inequality,* no? Correct: it's an inequality where nothing is more complicated than a quadratic – for example, $2x^2 + 3x > 2$. A cubic inequality involves your variable cubed, as in $2x^3 + x > 0$.

In quadratic and cubic inequalities, negative numbers become a serious problem: it's very difficult to come up with a simple algorithm for solving such inequalities. Instead, you need to – gasp! – understand what's going on. In this section, I give you two tried-and-true methods for getting to that stage: the graph-sketching method (which I think is best) and the change-of-sign method (which is a bit dull, if you ask me).

Graph-sketching method

My preferred method of working out where a quadratic inequality holds is to sketch a graph. If you have to find the set of values of x where the inequality $x^2 - 11x + 24 > 0$, here's what I suggest you do:

1. **Make a plan.**

   You're going to sketch $y = x^2 - 11x + 24$ and work out where the y-values are positive.

2. **First, solve the quadratic as if the > were an =.**

   The quadratic would factorise as $(x - 3)(x - 8) = 0$, so you know that the curve crosses the x-axis at $x = 3$ and $x = 8$.

3. **Work out the shape of the graph.**

   It's a positive quadratic, so it'll be a smiley-face graph.

4. **Sketch it!**

   It should look like the one in Figure 6-2.

5. **Find where the inequality is true.**

   The curve is above the x-axis when $x < 3$ or $x > 8$, which is your answer.

When you have two separate possibilities (an 'outie'), the answer should be $x < (\text{the lower value})$ or $x > (\text{the higher value})$; if you have an 'innie', it should be $(\text{lower}) < x < (\text{higher})$.

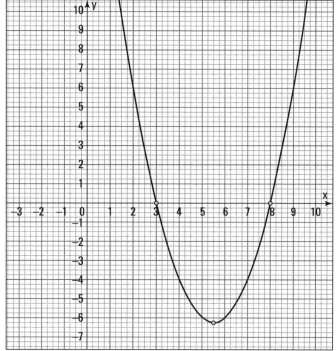

Illustration by John Wiley & Sons, Inc.

Figure 6-2:
Sketch of
$y = x^2 - 11x$
$+ 24$.

Change-of-signs method

The change-of-signs method makes use of the fact that every time a graph crosses the x-axis, the y-value changes from positive to negative or vice versa. If you can find out where $y = 0$, you can make a table to show where your inequality holds. The method looks like this:

1. **If you haven't already, find where $y = 0$ by factorising (or using the quadratic formula, if needed).**

   For the example from the preceding section, $x^2 - 11x + 24 = 0$, you have $x = 3$ or $x = 8$.

2. **Make a table with two rows; the number of columns should be one more than double the number of your solutions.**

   In this case, you need five columns. For a cubic with three solutions, you'd need seven.

3. **Put the solutions you've worked out in the even columns, leaving the odd columns blank.**

   Below the solutions, write 0, as in Figure 6-3a.

4. Pick a number below your first solution (say, 0), and put it into the function.

You get 24, which is positive, so write a + in the first blank space of the second row.

5. Pick a number between your solutions (say, 5), and put it into the function.

You get –6, which is negative, so write a – in the next blank space. (If you have more than two solutions, repeat this between each pair of solutions.)

6. Pick a number larger than your last solution (say, 10) and put it into the function.

You get 14, which is positive, so put a + in the last blank space. Your table should look like the one in Figure 6-3b.

7. Decide which regions satisfy your inequality.

You wanted positive answers, which you get when $x < 3$ or $x > 8$. That's the answer!

Figure 6-3:
A change-of-sign table (a) in progress and (b) finished.

(a)

x		3		8	
y		0		0	

(b)

x	< 3	3	3–8	8	> 8
y	+	0	–	0	+

Illustration by John Wiley & Sons, Inc.

Combining inequalities

A common variant on standard inequality questions involves combining two separate inequalities into one – for example, a playground may have a minimum area and a maximum perimeter, and you need to give the range of values that would satisfy both. My preferred approach to these problems is to use a number line.

To take a concrete example, say you've solved one inequality to show that $x > 5$ and another to show that either $2 < x < 8$ or $x > 12$. Here's what you'd do:

1. **Sketch out a number line and mark the numbers you've been given on it.**

   Don't bother measuring; it's a sketch. (See Figure 6-4a.)

2. **Draw your first solution set above the number line.**

   Here, you draw an open circle at 5 and a long arrow pointing to the right.

3. **Draw your second solution set above the number line, too.**

   Here, you draw open circles at 2 and 8, connected by a line, and you draw an open circle at 12 with a long arrow pointing to the right. (See Figure 6-4b.)

4. **You're interested in any intervals where you can see two lines above the number line.**

   There are two such intervals: between 5 and 8 and from 12 and up. Because you want to find the values of x that satisfy both solution sets, your answer is '$5 < x < 8$ or $x > 12$'.

Figure 6-4:
A number
line in
progress (a)
and com-
pleted (b).

(a)

(b)

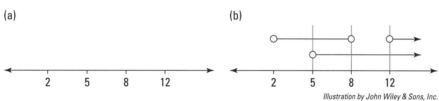

Illustration by John Wiley & Sons, Inc.

Discriminant-related inequalities

One of the most confusing things in Core 1 for many students is the type of question that says something like '$kx^2 + 6x + (4k + 5) = 0$, where k is a constant, has no real solutions. Find the set of possible values of k.'

Three things seem to frazzle people:

- ✔ There's a quadratic inequality.

- ✔ The inequality is about k rather than about x.

- ✔ The inequality can be either an 'innie' like $3 < k < 7$ or an 'outie' like $k < 3$ or $k > 7$.

There's no need to get frazzled. Breaking the problem down into steps makes it completely possible:

1. **Use the number of real roots to write down an inequality with the discriminant.**

   Here, $b^2 - 4ac < 0$, so $36 - 4k(4k+5) < 0$.

2. **Rearrange this into a nicer form.**

   Being *extremely* careful with the direction the inequality goes, you get $16k^2 + 20k - 36 > 0$, or $4k^2 + 5k - 9 > 0$.

3. **Factorise the quadratic and sketch the curve.**

   $(4k+9)(k-1) > 0$, and a sketch of $y = (4k+9)(k-1)$ is in Figure 6-5.

4. **Spot where the graph satisfies the inequality.**

   Here, you want the *y*-value to be positive – that is, you want the *k*-values that give you a *y* that's bigger than 0. If $k < -\dfrac{9}{4}$ or $k > 1$, you have a positive *y*. (This is an 'outie', so you need two separate inequalities.)

5. **Write down your final answer!**

   '$k < -\dfrac{9}{4}$ or $k > 1$'

WARNING!

Make sure your final answer is in terms of the correct letter – here, you're putting constraints on the constant *k*, not the variable *x*.

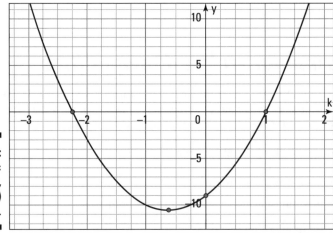

Figure 6-5:
A quadratic sketch, $y = (4k+9)(k-1)$.

Illustration by John Wiley & Sons, Inc.

Chapter 7

Factors, Remainders and Fractions

· ·

In This Chapter

▶ Identifying factors

▶ Using the remainder theorem

▶ Dividing out factors for fun

▶ Putting fractions in their simplest form

▶ Picking out partial fractions

· ·

*Y*ou did some factorising at GCSE – splitting numbers up into their prime factors (for example, $45 = 3^2 \times 5$), putting algebraic expressions into brackets ($x^2 + 3x = x(x+3)$ and $x^2 + 9x + 8 = (x+1)(x+8)$), and so on. Chapter 3 has a section devoted to factorising quadratics, and now would be an awesome time to refresh your memory on that, especially the ones with a number (other than 1) in front of the x^2.

Factorising is a good skill to have. Especially in Core 2, it will serve you well. You're going to take it a bit further, though; you'll see some ways of finding algebraic factors and of methodically finding the other factors after you have the first one.

What you're dividing doesn't always divide exactly, so this chapter also tells you how to deal with remainders, both in the simple Core 2 case and in Core 4, where things can be a bit more involved.

Lastly, dealing with fractions properly is one of the places I've seen many A level students come to grief, so I go over some of the skills you need to have in place to stay ahead of the curve. One of those skills is partial fraction decomposition, which can turn an apparently impossible integration problem into one that's just tricky.

Finding Factors

GCSE hints about the *factor theorem* but tells you no more. The main idea is that if you substitute a certain value (let's call it a) for x in a function or expression and the result is 0, then $(x-a)$ must be a factor – because substituting in $x = a$ clearly makes $(x-a)(\text{anything else})$ equal to 0.

This means that a is a *root* of the function or expression.

More formally, the Factor Theorem says: a polynomial $f(x)$ has a factor $(x-a)$ if and only if $f(a) = 0$.

More often than not, you'll be given a very strong hint about what might be a factor; for example, the question might subtly ask you to 'show that $(x-3)$ is a factor of $2x^3 - 7x^2 + 9$'. However, the problem isn't always so clear-cut.

When you aren't given a broad hint as to which factors might be lurking in a cubic function, you'll need to use sophisticated techniques like 'guessing' and 'guessing cleverly' to find them. More specifically, there are two methods I recommend: using your calculator's Table mode, and trial and improvement. Another option may be to use your calculator's FACT button. I describe all three approaches next.

Turning the tables

I hope you have a calculator with a Table mode. If not, you should definitely get one – I use a humble Casio FX-83GT, which does everything I need it to do; other calculators are available.

One of the things I need my calculator to do is to quickly work out a factor – and for that, Table mode is invaluable. Here's how to find a factor:

1. **Put your calculator in Table mode**.

   Press MODE (usually somewhere in the top right) and then whichever number has TABLE written beside it on the screen – on mine, that's 3.

2. **Type in the expression you're trying to factorise.**

   You'll find an x by pressing ALPHA in the top right and then whichever key has a red X above it – on mine, it's the close-bracket, ')'. After you have, say, $2x^3 - 7x^2 + 9$ typed in, press '='.

3. **Enter your domain.**

   Usually –5 to 5 is plenty. Type '-5' and then '=' when it says 'Start?'
   When it says 'End?', type 5 and then '='. Type 1 and then '=' when it asks
   'Step?', although you might try 0.5 and see what happens.

4. **Find your root in the table.**

   Up will pop, after a second or two of thinking time, a table listing values
   of x and of the function you asked the calculator to work out. Look for a 0
   in the last column; the corresponding number in the column before it is a
   root (a), and $(x-a)$ is a factor.

If you're lucky, you might get several factors this way!

Using trial and improvement

Trial and improvement is an old favourite from your GCSE days, right? You
can narrow down where a root is for a cubic (or any continuous curve) by
finding an x-value that gives you a negative result, another x-value that gives
you a positive result, and repeatedly bringing the x-values closer together.
Here's how, still with $2x^3 - 7x^2 + 9$:

1. **Pick an x-value that gives you a positive result and an x-value that
   gives you a negative result.**

   For this example, $x = 0$ gives you a positive result (9), and $x = -3$ gives
   you a negative result (–108).

2. **Pick the x-value midway between your current values, and work out
   the answer there.**

   When $x = -1.5$, you get –13.5, which is a negative result.

3. **One of your current x-values gives the same signed result as the midpoint you just worked out; replace it with the midpoint.**

   Here, you'd replace –3 with –1.5.

4. **Repeat Steps 2 and 3 until you converge on a root.**

   The next x-value would be –0.75, which gives you a positive answer and
   replaces 0, so now you have –1.5 and –0.75. Their midpoint is –1.125,
   which gives you a negative answer, so it replaces –1.5. Repeat this
   process for long enough, and you'll end up with $x \approx -1$, which tells you
   $(x+1)$ is likely a factor.

5. **Check your probable factor works using the factor theorem.**

   Try $x = -1$ in your original expression. If it's a root, your result should be
   0. It is. You've found a factor.

At A level, factors to be found this way are usually integers, frequently halves, and quite rarely other fractions. If you've reached a point where there's only one integer between your x-values, it's generally worth checking immediately whether that's a root. If not, you need to keep going!

The prime directive: Getting the FACTs

If your calculator has a FACT button, you're in luck! It gives you the prime factorisation of integers. There's a nice way to factorise (in principle) any polynomial expression using your calculator. Here's what you do, again using $2x^3 - 7x^2 + 9$:

1. Store the number 1,000 as X.

Type '1000', then STO (SHIFT and RCL in the middle on the left), and then the button with a red X above it (typically the close-bracket button).

2. Now type your polynomial in.

Use the ALPHA button to get an X.

3. Press '='.

You'll see a large integer (1,933,000,009 in this case), which is the value of the polynomial when $x = 1,000$. Don't worry about it.

4. Press FACT.

On my machine, you get FACT button by pressing SHIFT, then the button with commas. The calculator thinks for a moment and then gives you $7 \times 11 \times 13 \times 997 \times 1,997$.

5. See how the biggest number there relates to $x = 1,000$.

1,997 is pretty clearly $2x - 3$. Type in *Ans* $\div 1997$ to get a smaller number, 997,997. (You're effectively factoring out $2x - 3$ here.)

6. Keep doing the factorising!

You might spot here that $997,997 = 1,001 \times 997$ or $(x+1)(x-3)$, but even if not, pressing FACT again would give you the 997 straight off.

You end up with three factors, and $(2x-3)(x+1)(x-3)$ is indeed the fully factorised form of $2x^3 - 7x^2 + 9$.

This method sometimes gives you numbers that aren't obviously close to multiples of 1,000, like $3^3 \times 37$ (which happens to be 999). After a while, you get used to looking for these; there's almost always another number you *can* spot. If not, then you might want to try a different method.

The main reason for picking 1,000 as x is that it's quite easy to spot nearby factors. Any integer is perfectly fine, but some are easier to work with than others. If you find 100 (or, for that matter, something strange like 314) works better for you, go for it!

Rooting Out Remainders

You probably know about remainders from when you first learnt division – for example, $50 \div 8$ is 6 (remainder 2). A remainder is what's left over when you've taken out everything that can be divided out exactly, and it's what goes on top of the fractional part if you want that fraction (you could also write $50 \div 8 = 6 + \frac{2}{8}$ before cancelling it down).

In Core 2, you're often told that a given polynomial has a particular remainder when divided by something, or you're asked to find the remainder in a given situation.

When given this kind of remainder question, it's tempting to do long division to find the remainder. The same goes for factor-theorem questions. *Do not do this.* You might get the marks, eventually, but you'll waste an awful lot of time. It's much, much better to use the remainder or factor theorem directly.

The *remainder theorem* is a close cousin of the factor theorem, and it says the following: If $f(x)$ is a polynomial and the remainder when you divide $f(x)$ by $(x-a)$ is b, then $f(a) = b$ (and vice versa).

If you're asked to find the remainder of a polynomial when it's divided by a given expression, you just put whatever would make the expression 0 in as x, and what comes out is your remainder – much less work than doing the whole division!

For example, the remainder when you divide $3x^3 + 7x^2 - 4x + 3$ by $(x-1)$ is 9, because when you evaluate the expression with $x = 1$, it gives you a result of $3 + 7 - 4 + 3 = 9$.

You might also be given a polynomial with an unknown coefficient and be told it gives you a specific remainder when divided by something; then you're asked to find the unknown coefficient. For example, the polynomial could be $x^3 + 3x^2 + ax + 10$, which gives you a remainder of 6 when divided by $(x+2)$. Here's what you'd do:

1. Work out what you need to put into the polynomial.

Here, it's $x = -2$ that solves $x + 2 = 0$.

2. Substitute this value for x.

$$(-2)^3 + 3(-2)^2 + a(-2) + 10$$
$$= -8 + 12 - 2a + 10$$
$$= 14 - 2a$$

3. This has to be equal to the remainder – solve for a!

$14 - 2a = 6$, so $2a = 8$ and $a = 4$. That's the answer.

Dividing Out Factors

When you know something is a factor, there's only one thing a mathematician would ever do next: divide it out! It's like some sort of compulsion. Actually, it's because factorising appeals to a sense of mathematical neatness and tells you a lot more about a function than an expanded form.

A factorised function is a happy function!

In this section, I give you two methods for dividing a factor out of an expression – first, the traditional *long division* method you're likely to have seen in school, and then a much less-error-prone equivalent method called *matching coefficients*.

Long (and tedious) division

Let's say that you're told $f(x) = 3x^3 + 2x^2 - 19x + 6$ and that you've shown (using the factor theorem) that $(x-2)$ is a factor. (If you want to check that $f(2) = 0$, please go ahead!) The question will inevitably then ask you to factorise $f(x)$ fully or to give the solutions for $f(x) = 0$, which amounts to much the same thing.

You're going to need a bus stop, set out like in Figure 7-1. Follow along!

1. Write your function as descending powers of x.

In this case, you don't need to do anything – it's already going down from cubes to squares to xs to constants. But if you were missing, say, an x^2 term, you'd want to add $0x^2$ in the appropriate place. Give yourself plenty of space.

2. **Draw a bus stop over the function you're trying to divide, and write what you're dividing by in front.**

   See Figure 7-1a to make sure you have it right.

3. **Look at the first term under the bus stop and the first term of the thing you're dividing by, and divide them. Write the answer above the bus stop.**

   Here, $3x^3 \div x = 3x^2$, so this goes on top.

4. **Multiply what you just wrote down by the thing you're dividing by and write it underneath.**

   $3x^2(x-2) = 3x^3 - 6x^2$, as in Figure 7-1b.

5. **Take this away, very carefully, from what's under the bus stop.**

   You get $8x^2 - 19x + 6$.

6. **Repeat Steps 3 through 5 until you're left with either 0 or a polynomial of smaller degree than what you're dividing by.**

   In this case, the next thing to go up is $8x$, and you take off $8x(x-2) = 8x^2 - 16x$ to get $-3x + 6$. Finally, -3 goes up, and you take off $-3(x-2) = -3x + 6$ to end up with no remainder.

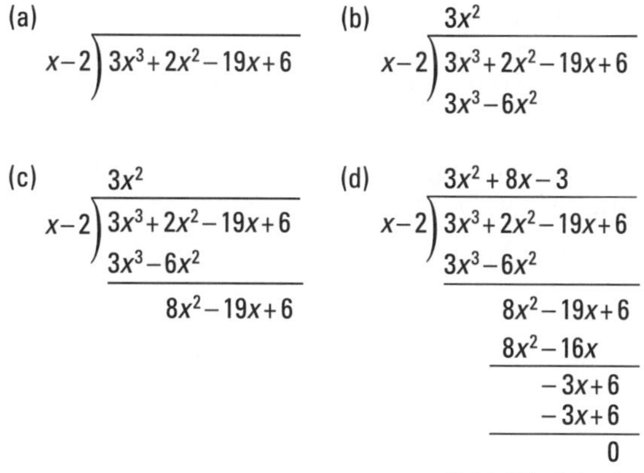

(a)
$$x-2 \overline{\smash{)}\, 3x^3 + 2x^2 - 19x + 6}$$

(b)
$$\begin{array}{r} 3x^2 \\ x-2 \overline{\smash{)}\, 3x^3 + 2x^2 - 19x + 6} \\ 3x^3 - 6x^2 \end{array}$$

Figure 7-1:
Polynomial
long
division: a)
initial setup;
b) and c) first
iteration;
d) final
result.

(c)
$$\begin{array}{r} 3x^2 \\ x-2 \overline{\smash{)}\, 3x^3 + 2x^2 - 19x + 6} \\ 3x^3 - 6x^2 \\ \hline 8x^2 - 19x + 6 \end{array}$$

(d)
$$\begin{array}{r} 3x^2 + 8x - 3 \\ x-2 \overline{\smash{)}\, 3x^3 + 2x^2 - 19x + 6} \\ 3x^3 - 6x^2 \\ \hline 8x^2 - 19x + 6 \\ 8x^2 - 16x \\ \hline -3x + 6 \\ -3x + 6 \\ \hline 0 \end{array}$$

© John Wiley & Sons, Inc.

The expression at the top, $3x^2 + 8x - 3$, is the result of dividing $3x^3 + 2x^2 - 19x + 6$ by $x - 2$. (You're itching to factorise that, aren't you? Good. Go ahead! I'll catch up with you in a few sections' time.)

If you're dividing by something you know is a factor and end up with a remainder, something has gone wrong. Most likely you've lost a minus sign somewhere. It's easily done.

Surely there must be a better way? Of course there is. Read on.

Matching coefficients

The method I recommend is called *matching coefficients* – I think what's going on is a bit more obvious.

The trick is to turn the question from a difficult division into a much simpler multiplication – and figure out what you have to multiply by! With the same example as before, here's how it works:

1. **Decide which degree of polynomial you're expecting.**

   If you're dividing a cubic (degree 3) by a linear expression (degree 1), you expect a quadratic (degree $3 - 1 = 2$) as your answer.

2. **Decide what kind of remainder you're expecting.**

   If you know you're working with a factor, there's no remainder; otherwise, the remainder will be one degree less than what you're dividing by. For example, if you're dividing by a linear expression, you'll get a constant remainder, and dividing by a quadratic would give you a linear remainder.

3. **Write (what you're dividing) = (what you're dividing by)(answer you're expecting) + (remainder you're expecting).**

   Here, you're expecting some quadratic as an answer and no remainder, so

   $$3x^3 + 2x^2 - 19x + 6 = (x - 2)(ax^2 + bx + c)$$

4. **Multiply out and simplify the right-hand side.**

   $$ax^3 + (b - 2a)x^2 + (c - 2b)x - 2c$$

5. **Match up the coefficients!**

   You need the same number of x^3s on each side, so a has to be 3. You need the same number of x^2s, too: $2 = b - 2a$. But you know a is 3, so $b = 8$. Lastly, you know $-19 = c - 2b$, and b is 8, so $c = -3$!

It's a good idea to check with the last remaining equation: is $-2c$ equal to 6? Yes, it is, which is reassuring!

I'd lay this out as follows – it's a bit easier to keep track of what's going on.

$$3x^3 + 2x^2 - 19x + 6 = (x-2)\,(ax^2 + bx + c)$$
$$3x^3 + 2x^2 - 19x + 6 = ax^3 + (b-2a)\,x^2 + (c-2b)\,x - 2c$$

$$3x^3 = ax^3$$
$$2x^2 = (b-2a)\,x^2$$
$$-19x = (c-2b)\,x$$
$$6 = -2c$$

That means your quadratic factor is $3x^2 + 8x - 3$, like in the preceding section.

Finishing off the question

This step is important enough that it has its own heading, not just a Remember or Warning icon: *don't forget to finish the question.*

You're not finished when you figure out the factors. You're not even finished when you factorise the factor you found (here, $3x^2 + 8x - 3$ becomes $(3x-1)(x+3)$). You're done when you answer the question they've asked.

> ✔ If they've asked you for the fully factorised form of $f(x)$, give them the fully factorised form; $f(x) = 3x^3 + 2x^2 - 19x + 6$ becomes
>
> $$f(x) = (3x-1)(x+3)(x-2)$$
>
> ✔ If they've asked you for the solutions where $f(x) = 0$, give them the solutions where $f(x) = 0$. Here, you'd write '$x = \frac{1}{3}, x = -3$ or $x = 2$'.
>
> ✔ If they've asked for something else, give them the thing they ask for.

I've said many times – and will continue to say – that good mathematicians are lazy, and when you've got your secret mathematician card with all the digits of π on it, you can leave this sort of thing unfinished. Until then, make sure you give the examiners exactly what they ask for. Don't throw away marks!

Dealing with remainders

If you have a remainder in a division problem involving whole numbers – like $100 \div 3 = 33$ with 1 left over – you handle the remainder by putting it on top of a fraction, the bottom of which is what you divided by: $100 \div 3 = 33\frac{1}{3}$.

I'm sure you'll be stunned to hear that the same thing goes for algebraic division. If you divide $x^2 + 3x + 2$ by $x - 1$, you get $x + 4$ and a remainder of 6. That means $\dfrac{x^2 + 3x + 2}{x - 1} = x + 4 + \dfrac{6}{x - 1}$.

In Core 4, if you're dividing by a quadratic, you can sometimes split the remainder up into partial fractions, which live later in this chapter.

Putting the Factor and Remainder Theorems Together

A full-on factor and remainder theorem question might go something like this – feel free to work it through on your own before I give you my solution.

$f(x) = 4x^3 - 8x^2 + ax + b$, where a and b are constants.

a) When $f(x)$ is divided by $(x - 1)$, the remainder is 14. When $f(x)$ is divided by $(x + 1)$, the remainder is 60. Find a and b.

b) Show that $(x - 3)$ is a factor of $f(x)$.

c) Find all solutions of $f(x) = 0$.

You could expect around six marks for part a), two for part b) and maybe three for part c).

Solving part a)

For part a), the only sensible way to approach it is to use the remainder theorem, which says that, in this case, $f(1) = 14$ and $f(-1) = 60$. Work those out for $f(x) = 4x^3 - 8x^2 + ax + b$:

$$f(1) = 4 - 8 + a + b = 14, \text{ so } a + b = 18$$

$$f(-1) = -4 - 8 - a + b = 60, \text{ so } b - a = 72$$

Using whichever simultaneous-equations method suits you best (see Chapter 3), you can figure out that $b = 45$ and $a = -27$. Those are unusually large numbers for this type of question, so you should treat them with suspicion; luckily, if they're wrong, part b) won't work, so you've got a check built in.

Solving part b)

Part b) asks you to show that $(x-3)$ is a factor of $f(x)$; you've just found that $a = -27$ and $b = 45$, so the function is $f(x) = 4x^3 - 8x^2 - 27x + 45$. This part is easily solved using the factor theorem: simply evaluate $f(3)$. This gives you $4 \times 27 - 8 \times 9 - 27 \times 3 + 45 = 108 - 72 - 81 + 45 = 153 - 153 = 0$, which means $(x-3)$ is a factor of $f(x)$.

Solving part c)

Part c), which asks for all solutions of $f(x) = 0$, is more tedious. You know $(x-3)$ is a factor of the cubic $4x^3 - 8x^2 - 27x + 45$, so you can write that as $(x-3)(ax^2 + bx + c)$. Multiplying out gives you

$$4x^3 - 8x^2 - 27x + 45 = ax^3 + (b - 3a)x^2 + (c - 3b)x - 3c$$

Now you can match coefficients:

$$4x^3 = ax^3$$
$$-8x^2 = (b - 3a)x^2$$
$$-27x = (c - 3b)x$$
$$45 = -3c$$

So $a = 4$ directly, and $c = -15$; if $b - 3a = -8$, then $b = 4$. Just check that $c - 3b = -27$, which it does. Great! The quadratic factor is $4x^2 + 4x - 15$, which factorises as $(2x + 5)(2x - 3)$.

That means $f(x) = (2x + 5)(2x - 3)(x - 3)$, so if $f(x) = 0$, x is $-\frac{5}{2}, \frac{3}{2}$ or 3.

Simplifying Fractions

It's not at all uncommon for the start of a Core 3 exam to try to lead you into the test gently by asking you to simplify a fraction like this:

$$\frac{x+2}{4x^2 - 16} - \frac{1}{4x + 1}$$

It's also not at all uncommon for students to throw a wobbly when they see such a thing. There's absolutely no reason to, though – in principle, you should have been dealing with these at GCSE.

Here's what you'd do:

1. **Put everything into brackets and see if there's simplifying to be done.**

   The bottom of the first fraction factorises as $4(x-2)(x+2)$, and you can cancel the $(x+2)$s on the top and bottom to leave

   $$\frac{1}{4(x-2)} - \frac{1}{4x+1}$$

2. **Now take away the fractions as normal.**

   Get a common denominator and subtract:

   $$\frac{(4x+1)-4(x-2)}{4(x-2)(4x+1)}$$

   The top becomes $4x+1-4x+8 = 9$, leaving you with $\dfrac{9}{4(x-2)(4x+1)}$, and nothing more cancels.

A fraction in brackets is a happy fraction.

Apart from slips like cancelling things that ought not to be cancelled, the biggest rabbit-holes I've seen students falling down (after they've stopped with the wobbly-throwing) are not finding the common denominator first and not leaving things in their simplest form at the end.

Cancelling first to get lowest common denominators

It's always tempting to fly at a fractions question with rules rather than thought. That sometimes works, but it sometimes leaves you in a mess, too. For example, a Core 3 question might ask you to simplify $\dfrac{2(x-1)}{x^2-2x-3} - \dfrac{1}{x-3}$, and I've seen students use their usual 'multiply the bottoms together, then...' algorithm to end up with a quadratic on top and a cubic on the bottom, possibly $\dfrac{x^2-6x+9}{x^3-5x^2+3x+9}$ if they're very careful with their signs, and then say, 'What now?' The answer to that, incidentally, would be 'factorise'. But there's a better way, which is as follows:

1. **Factorise as much as possible first.**

   The first fraction is $\dfrac{2(x-1)}{(x-3)(x+1)}$.

2. Find a common denominator.

Because the first and second fractions share the $(x-3)$ on the bottom, $(x-3)(x+1)$ makes a great common denominator.

3. Make sure both fractions have the same bottom.

Multiply the second one, top and bottom, by $(x+1)$ to make it

$$\frac{x+1}{(x-3)(x+1)}$$

4. Do the sum.

The top becomes $2(x-1)-(x+1)=x-3$, and the bottom is $(x-3)(x+1)$.

5. Simplify!

There's an $(x-3)$ on the top and the bottom, so cancel those, giving you a final answer of $\frac{1}{x+1}$.

Moral of the story: simplifying up front can save you a lot of effort.

Understanding simplest form

It's not always clear what constitutes the 'simplest form' of a fraction – even to me, and I'm supposed to be an expert. I imagine that if you asked three or four different mathematicians what the simplest form of a nasty algebraic fraction was, you'd get three or four different answers. What I *can* do is give you some guidelines about what I understand 'simplest form' to mean, without injecting too many of my own personal prejudices; these represent a sort of consensus view of simplest form. A lowest common denominator, if you like.

- ✔ Unless you're doing partial fractions, you should end up with only one fraction.

- ✔ Both the top and bottom of the fraction should be fully factorised. A factorised fraction – as I may have mentioned – is a happy fraction.

- ✔ If the same bracket shows up twice anywhere, the chances are there's either more factorising to do or something you can cancel.

- ✔ Avoid decimals. If you have decimals, multiply top and bottom by some power of ten and get rid of them.

- ✔ Avoid stacked fractions. If you have a fraction on the top or bottom, multiply top and bottom by the bottom and repeat until the stacked fractions are all gone.

✔ Use trig identities where needed – you don't want $\dfrac{\cot(x)}{\csc(x)}$ when you could have $\cos(x)$ instead.

✔ Prefer to keep powers positive in a fraction – it's better to have e^x on the bottom than e^{-x} on the top.

✔ Mildly prefer minor trig functions on top over regular ones on the bottom. I'd rather see $\tan(x)\sec(x)$ than $\dfrac{\tan(x)}{\cos(x)}$ or even $\dfrac{\sin(x)}{\cos^2(x)}$.

Piecing Together Partial Fractions

You are – or at least I hope you are – happy with the idea of taking $\frac{1}{3} + \frac{1}{4}$ and turning it into $\dfrac{3+4}{(3)(4)} = \dfrac{7}{12}$. Similarly, you can take $\dfrac{3}{x+2} + \dfrac{4}{x-3}$ and make it $\dfrac{3(x-3)+4(x+2)}{(x-3)(x+2)} = \dfrac{7x-1}{(x-3)(x+2)}$. Partial fractions reverse this process – you start with a single fraction whose bottom is complicated and turn it into two or more fractions with simpler bottoms.

Why is this useful? Mainly for integration. You see in Chapter 16 how to integrate simple fractions – but the method for complicated fractions is usually to turn them into simple fractions!

You may also use partial fractions to turn a messy fraction into something you can apply the binomial expansion to – that's in Chapter 8 if you need it.

The basic method

About 90 per cent of partial-fractions questions either will have a denominator without squares in it or will give you a template of what the answer ought to look like. The template usually looks something like this:

$$\frac{3x^2+1}{(x+1)^2(x-3)} = \frac{A}{x+1} + \frac{B}{(x+1)^2} + \frac{C}{x-3}$$

That is, a load of letters on top of fractions.

If you've got something with a square in and no template, tough luck: you need to go to the later section 'Trickier denominators'. If you're given a template, though, you can skip the next subsection and go straight to the method; if not, read on!

Making your own template

If they've not given you a template, you may shake an angry fist in their direction. It won't help you solve the problem, but it might make you feel better. After that, you should follow this recipe. And so you have something to work with, let's sort out $\dfrac{10x^2 - 21x - 6}{2x^3 - 5x^2 - 3x}$:

1. **Factorise the bottom of the fraction you're working with.**

   Here, you have $x(x-3)(2x+1)$; all the factors are linear, which is nice. See Chapter 6 if you need help with factorising.

2. **Take each factor and put it on the bottom of a fraction, each of which should have a different letter on top, with plus signs in between.**

   For example, $\dfrac{10x^2 - 21x - 6}{x(x-3)(2x+1)} \equiv \dfrac{A}{x} + \dfrac{B}{x-3} + \dfrac{C}{2x+1}$. That's your template!

A common question is 'Why are they added rather than taken away?' The answer is, it doesn't *really* matter – a plus is normally easier to work with than a minus, and if it should have been a minus, you'll just end up with negative constants. It all comes out in the wash!

Working out the constants

To work out the missing constants in your partial-fractions template, here's what you do:

1. **Multiply both sides by the denominator from the left-hand side, cancelling where possible.**

   Your template is

   $$\frac{10x^2 - 21x - 6}{x(x-3)(2x+1)} \equiv \frac{A}{x} + \frac{B}{x-3} + \frac{C}{2x+1}$$

   In this example, you get

   $$10x^2 - 21x - 6 \equiv A(x-3)(2x+1) + Bx(2x+1) + Cx(x-3)$$

2. **Pick a value of x – preferably one that 'disappears' at least one constant.**

   For instance, substituting $x = 0$ into both sides gives you $-6 = A(-3)(1) + B(0)(1) + C(0)(-3)$, or $-6 = -3A$; A is 2. In general, any value that makes the bottom of one of the original fractions 0 is usually an excellent choice for x. Any value of x will work, but some values are easier to work with than others!

3. Repeat until you have all the constants.

Use $x = 3$ to get rid of A and C, leaving you with $21 = 21B$, so $B = 1$.

You can then use $x = -\frac{1}{2}$ to get rid of A and B. That leaves you with $7 = \left(-\frac{1}{2}\right)\left(-\frac{7}{2}\right)C$, making $C = 4$.

4. Write out the final result.

$$\frac{10x^2 - 21x - 6}{x(x-3)(2x+1)} \equiv \frac{2}{x} + \frac{1}{x-3} + \frac{4}{2x+1}$$

There you have it: your one horrible fraction split up into three (comparatively) nice ones.

Trickier denominators

When writing partial fractions, the trouble comes when you have a factor squared or, heaven forfend, cubed on the bottom. (You don't have to worry about cubes at A level, thankfully).

If you have a whole linear bracket squared on the bottom, like in $\frac{4x+5}{(x+1)(x-2)^2}$, you can't get away with a template just involving $\frac{A}{x+1}$, $\frac{B}{x-2}$ and $\frac{C}{x-2}$ – sadly, the bottoms all need to be different. The way around it is to square the bottom of the C fraction, so your template would be $\frac{A}{x+1} + \frac{B}{x-2} + \frac{C}{(x-2)^2}$.

In some boards, you also need to deal with things like $\frac{3x-1}{(x^2+1)(x-1)}$. You'd think that $\frac{A}{x^2+1} + \frac{B}{x-1}$ would be a perfectly good template, but no! Booby-trap. Above a quadratic expression, you need to have a linear expression. If you had a cubic on the bottom, you'd need a quadratic on top. In general, the biggest power in the polynomial on the top needs to be one less than the biggest power on the bottom.

So, above x^2+1, you need to have a *linear* expression such as $Ax+B$. In this case, your template would be something like $\frac{Ax+B}{x^2+1} + \frac{C}{x-1}$.

Chapter 8

Getting Serious about Series

• •

In This Chapter

▶ Defining series explicitly and recursively

▶ Analysing arithmetic and geometric series

▶ Breaking down the binomial expansion

• •

*Y*ou may have seen a few sequences at GCSE – there's the occasional throwaway question about finding the *n*th term of an arithmetic or quadratic sequence, but there's very little by way of exploring the rich and incredibly deep ideas you can express using series.

So what is a series? Strictly speaking, there's a difference between a sequence and a series:

✔ A *sequence* (also called a *progression*) is an ordered list of *terms* (usually numbers, but generally 'things you can add up') such as 4, 6, 8, 10, . . .

✔ A *series* is what you get if you add them all up: $4 + 6 + 8 + 10 + \ldots$

However, the words are quite often used interchangeably in casual speech. That means 'I get the correct words mixed up and nobody ever picks up on it, but it doesn't mean you should.'

Mathematicians naturally like to give things very short names, and the terms of a sequence are often denoted by u_n. Why *u* should stand for 'the term of the sequence' is a mystery; there's nothing special about *u*, and it's perfectly acceptable for the terms of a sequence to be called v_n or a_i or anything else.

You may not have seen this *subscript* notation before – writing a small number or letter below and after a letter. It's nothing to do with powers; the subscript (often *n*, sometimes *k* or *i*) refers to the position of the term in the sequence: u_3 would be the third term, and u_{10}, the tenth.

The definition of a sequence should always include an inequality, something along the lines of $n \geq 1$. This is nothing to worry about: all it means is that the sequence is defined only when *n* is 1 or greater. You can't put –4 or 0 in – it doesn't make sense to have the –4th term of a sequence. (Also, n has to be an integer: it doesn't make sense to talk about the πth term of a sequence, either!)

I like to think of the subscript as a football shirt number and the main letter as the team name – the squad of the mighty champions u_n consists of the star players u_1, u_2, u_3 and so on!

Nothing says that the terms of a sequence have to be numbers – they can, in principle, be any mathematical object, but as far as you're concerned, they will always be either numbers or relatively simple expressions, such as multiples of a power of x.

In this chapter, I show you the various kinds of series and sequences you can expect to see in your A level. Depending on your board, you will encounter recursively and explicitly defined series and arithmetic series in either Core 1 or Core 2, whereas geometric series and the simpler kind of binomial expansion are always Core 2. In Core 4, you come across the nastier binomial type – but luckily, I have a shortcut method for you!

Explicit and Recursive Definitions

Explicit and recursive definitions are, simultaneously, some of the simplest ideas you'll see at A level and some of the most intimidating. They're a little intimidating because the notation is unfamiliar, and how to unpack it isn't immediately obvious. (Also, if your handwriting is as dicey as mine, the difference between u_{n+1} and u_n+1 may not be perfectly clear.)

In this section, I try to make the techniques for finding the nth term of a sequence or series as simple as possible – as soon as you get hold of the logic, this kind of Core 1 question offers some relatively easy marks!

Explaining explicitly defined sequences

An *explicitly defined* sequence is nothing to do with being rude – far from it! It's about the nicest sort of sequence you can hope to see. Here's an example:

$$u_n = n^2 + 3n + 2, \quad n \geq 1$$

This means that if you want to know the nth term in the sequence, you square n, add on 3 times n, and add 3. The first term is $1+3+2=6$, the second term is $4+6+2=12$, the tenth term is $100+30+2=132$, and so on.

An explicitly defined sequence gives you a precise recipe for finding any term in the sequence, without reference to any other term.

Getting your head around recursive sequences

A *recursively defined* sequence is one where each term refers to the one before it. For example:

$$u_1 = 2$$

$$u_{n+1} = 4u_n + 3, \quad n \geq 1$$

This takes a little bit of unpacking. The first line is pretty clear: it tells you the value of the first term. The second line is a bit more complicated; it's simplest to figure it out if you pick a valid value for n as given by the inequality, such as 1. In that case, you have $u_2 = 4u_1 + 3$. However, you know what u_1 is from before: it's 2, so $u_2 = 4(2) + 3 = 11$. Similarly, $u_3 = 4u_2 + 3$, which is $4(11) + 3 = 47$. To get each new term, you multiply the one before it by 4 and add 3. In general, here's the recipe to start writing down the first few terms of a recursively defined sequence:

1. **You're always given the first term; you can simply write it down.**

   Always write both sides of the definition down – it makes things easier to refer to later. Don't just write '2'; write '$u_1 = 2$'.

2. **To get the second term, replace every n in the complicated line with 1, and replace every u_1 with the value you already worked out. Calculate the result.**

3. **To get the third term, replace every n in the complicated line with 2, and replace any u_1 or u_2 with the appropriate value you've already worked out. Calculate the result again.**

4. **Keep going for as many terms as you're asked about! At each step, replace each of the ns with the appropriate number and work out the result using what you've worked out already.**

It's common for this kind of question to involve unknown constants and to ask you to work out their possible values at the end. For example:

$$u_1 = 3$$

$$u_{n+1} = au_n + n, \quad n \geq 1$$

You may be told that $u_3 = 16$ and be asked to find the possible values of a.

Step-by-step, you would do the following:

1. Write down u_1.

It's $u_1 = 3$.

2. Work out u_2.

You have $u_2 = au_1 + 1 = 3a + 1$.

3. Work out u_3.

You have $u_3 = au_2 + 2 = a(3a + 1) + 2 = 3a^2 + a + 2$.

4. Knowing the value of u_3, you can equate it to your expression.

Set the expression for u_3 equal to 16: $3a^2 + a + 2 = 16$.

5. Simplify and solve the quadratic.

$3a^2 + a - 14 = 0$, which factorises as $(3a + 7)(a - 2) = 0$, giving you $a = -\dfrac{7}{3}$ or $a = 2$.

6. Check if there were any restrictions on your constant. (For example, are you told it's positive?)

Here, there are none, so both of your answers are valid.

Series Stuff: Summing Up Sigma Notation

As if subscript notation weren't enough, you also need to deal with *sigma* notation. This is another example of mathematicians being commendably lazy in the way they write things, and it's best illustrated by an example. Instead of writing $1 + 2 + 3 + 4 + 5 + \ldots + 100$, a mathematician would most likely write the following:

$$\sum_{n=1}^{100} n$$

How's that for parsimony? The funny E (it's really a Greek letter S, or *sigma*, which stands for *sum*) means you have to add up the values of the thing that comes after it – if you see a sigma, you're automatically dealing with a *series* rather than a sequence. You start counting at whatever is at the bottom of the sigma (here, you're told n starts at 1) and end at the top (100). So this terse expression means 'Add up every n you get if you start counting at $n = 1$ and keep going until you reach $n = 100$.'

You might be asked to evaluate something like that – although if you are, the chances are it'll be an arithmetic or geometric series (on which you can find more later in this chapter), and you have formulas to help you there.

More commonly, though, you're given an explicitly or recursively defined series and are told to find the sum of the first few terms (which may look like $\sum_{n=1}^{4} u_n$, if you were after the first four terms). Or possibly you're given the sum of the first few terms and are told to find the value(s) of a constant. If you're given $u_1 = 4$, $u_n = 2u_{n-1} - a$, $n > 1$ and $\sum_{n=1}^{4} u_n = 16$, here's how you'd work out a:

1. **Write out as many terms of the series as you're asked for.**

   The series is $\sum_{n=1}^{4} u_n$, so you need the terms u_1, u_2, u_3 and u_4, replacing the n subscript with the numbers 1 (at the bottom of the sigma) through 4 (at the top). $u_1 = 4$, $u_2 = 8 - a$, $u_3 = 16 - 3a$ and $u_4 = 32 - 7a$.

2. **Add them up.**

   These sum to $60 - 11a$.

3. **Set up an equation with the expression you worked out in Step 2 equal to the sum value you're given.**

   You write down $60 - 11a = 16$.

4. **Solve the equation!**

   You find $44 = 11a$, so $a = 4$.

5. **Check to see whether there are any restrictions on the constant you've found and whether your answer satisfies them.**

   There are no restrictions given here, so your answer is fine.

Analysing Arithmetic Sequences

You've seen plenty of *arithmetic sequences* in your mathematical career: probably the first sequence you ever saw was an arithmetic progression. It went 1, 2, 3, 4, . . . Ah, the nostalgia. 'But hang on', some readers with astronaut parents will say. 'The first sequence *I* met was 5, 4, 3, 2, 1.' Others might have been taken on a protest march where 2, 4, 6, 8,... was the order of the day. These are also arithmetic progressions. So what does that mean?

An arithmetic sequence is anything where the gap between successive terms – the *common difference* – is always the same. For 1, 2, 3, 4, 5, the common difference is 1. For 5, 4, 3, 2, 1, it's –1. And 2, 4, 6, 8 has a common difference of 2.

In this section, I show you how to find any term of an arithmetic sequence, how to add up any number of terms, how to work out the formula for a given arithmetic sequence, and how to find anything else they might ask on the subject.

Finding a term

In the formula book, you're given two arithmetic sequence formulas, the first of which looks something like this:

$$u_n = a + (n-1)d$$

I object to this – and, in fact, to all the sequence formulas given – in the strongest possible terms. Although mathematics *has* conventions, you can't assume that everybody else uses the same ones as you. There's a paragraph missing here, which should tell you the following:

✔ That u_n is the *n*th term of the arithmetic sequence

✔ That a is the first term of the sequence (or u_1)

✔ That d is the common difference of the sequence

Failure to define terms is a big mathematical no-no, and the exam boards should be ashamed of themselves. However, I'd expect that the chances of their changing that are about the same as the odds of one of their 'real life' problems ever being the slightest bit realistic, so you'll have to remember what all the letters mean.

To find the *n*th term of an arithmetic sequence, all you need to do is plug numbers into the formula. For example, if you know *a* is 10 and *d* is 2.5, the eighth term will be

$$u_8 = 10 + (8-1) \times 2.5$$
$$= 10 + 17.5$$
$$= 27.5$$

You may wonder where the arithmetic sequence formula comes from. It's not especially difficult: each term is d more than the one before, and to get from the first term, $a = u_1$, to the nth term, u_n, requires $(n-1)$ steps. That means you're adding $(n-1)$ lots of d onto the first term, giving you $a + (n-1)d$.

Finding a sum

You're given two versions of the 'sum to n terms' formula in the book. They're equivalent but don't look quite the same. The first, using L to represent the last (nth) term in the series, is as follows:

$$S_n = \frac{1}{2}n(a+L)$$

The second is this:

$$S_n = \frac{1}{2}n(2a+(n-1)d)$$

You can convince yourself they're identical by replacing L in the first equation with $a + (n-1)d$; it's generally up to you which you find easier to use when working on a question. Personally, if I know a, d and n, I use the second; if I know a, L and n, I use the first. If I know them all, then I find the first is harder for me to mess up – you're entitled to a different preference!

Say you're to find the sum of all of the multiples of 3 between 100 and 200. A little bit of thinking suggests that if you wrote down the sum explicitly, you'd have $102 + 105 + 108 + \ldots + 195 + 198$. From now on, it's all formula work:

1. **Work out a.**

   It's 102.

2. **Work out either d or L.**

   In this case, $d = 3$ and $L = 198$.

3. **Work out n.**

   Making a mistake with this step is easy. I like to use $u_n = a + (n-1)d$ to say $198 = 102 + 3(n-1)$, which you can solve: $n = 33$.

4. **To work out the answer, use whichever sum formula you prefer.**

   Using the L version, you get $\frac{1}{2} \times 33 \times (102 + 198) = \frac{1}{2} \times 33 \times 300 = 4,950$.

   Using the d version, you get $\frac{1}{2} \times 33 \times (204 + 96) = \frac{1}{2} \times 33 \times 300 = 4,950$.
   Reassuringly, they give you the same answer.

Don't be afraid to write out the sums you're doing in full – especially if you have something with potential pitfalls like 33×150. In principle, that's something you should have no problem with, but exam pressure does funny things to people.

Finding parameters

Many arithmetic sequence and series questions give you a (the first term) and d (the common difference) and then ask you to make progress from there. However, it's also quite common for an exam to give you details about either the terms themselves or the sum to a certain number of terms and expect you to work backwards to find the first term and common difference – which you can then use to solve other questions.

There are many ways they can ask this sort of question, so I give some general guidance here after a specific example:

> The 10th term of an arithmetic series is 50, and the sum to 20 terms is 1,060. Find the sum to 15 terms.

1. **Translate the information you have into equations.**

   Here, you use the nth term equation, $u_n = a + (n-1)d$, to get $50 = a + (10-1)d$, and you use the sum equation, $S_n = \frac{1}{2}n(2a + (n-1)d)$, to get $1,060 = \frac{1}{2} \times 20(2a + (20-1)d)$.

2. **Simplify your equations so you can solve them.**

   $50 = a + 9d$ for the first equation, and $20a + 190d = 1,060$ for the second. You might even divide the second by 10 to get $2a + 19d = 106$.

3. **Solve the simultaneous equations that come out.**

   You get $d = 6$ and $a = -4$.

4. **Use this information to answer the question they ask.**

   The sum to 15 terms is

   $$\frac{1}{2} \times 15(2 \times (-4) + 14 \times 6)$$
   $$= \frac{15}{2}(-8 + 84)$$
   $$= 15 \times 38$$
   $$= 570$$

Finding n

Finding the number of terms, n, in a given series or sequence is generally as simple as filling in the right details in the right formula and solving it. If you want to find how many terms are in the sequence 10, 11, 12, . . . , 1,000, you do the following:

1. **Write down a and d.**

   They're 10 and 1, respectively.

2. **Write down the formula for the nth term of a sequence: $u_n = a + (n-1)d$.**

   Your strategy is to substitute everything you know into this and work out the n value corresponding to the last term.

3. **Substitute in the known values.**

   You know that the last term, u_n, has the value of 1,000. The first term, a, is 10, and d is just 1, leaving you with

   $$1,000 = 10 + (n-1)$$

4. **Solve.**

   $$n = 991$$

It's more common to see a question where you're given the sum and need to find n – which generally leads to a quadratic. For example, if you know an arithmetic series has a first term of 100 and a common difference of –2, after how many terms is the sum 910?

1. **Write down a and d.**

   Here, you're given them: 100 and –2, respectively.

2. **Write down the formula for the sum to n terms.**

   Your strategy for this is the same as before: fill in the stuff you know and solve for what you don't know (the number of terms, n.) Your formula here is

   $$S_n = \frac{n}{2}\left(2a + (n-1)d\right)$$

3. **Fill in the numbers you know.**

   $$910 = \frac{n}{2}\left(2 \times 100 - 2(n-1)\right)$$

   You can (and should) simplify this to $910 = n(101 - n)$, or $n^2 - 101n + 910 = 0$.

4. **Solve.**

   This factorises as $(n-91)(n-10) = 0$, so either $n = 10$ or $n = 91$.

5. **Check whether the context of the question makes either of the answers invalid.**

   Here, both answers work. But the question could ask for the first time the sum becomes 910, in which case the answer would be $n = 10$.

   Also, real-world considerations will prevent n from ever becoming negative. More likely, one of the answers will be negative, and you can't have a negative number of terms.

Generating Geometric Sequences

Here are some examples of geometric sequences:

- ✔ 16, 8, 4, 2, 1, . . .
- ✔ 1, 3, 9, 27, 81, . . .
- ✔ 5, $\frac{10}{3}$, $\frac{20}{9}$, $\frac{40}{27}$, . . .

In each case, you get from one term to the next by multiplying by a certain *common ratio* (*r*) – in the first one, each term is half the one before (so the common ratio is $\frac{1}{2}$); for the second, the ratio is 3; and for the last, it's $\frac{2}{3}$. Each of them (just like the arithmetic sequences earlier in this chapter) has a first term (*a*) – the first terms are 16, 1 and 5, respectively.

In this section, I show you how to do all the things you need to do with geometric sequences: find a given term, find the sum of the first n terms (or even to infinity, in some cases), find the parameters a and r when given information about the sequence, and put it all together in the kind of awful question they sometimes throw at you.

Finding a term

The formula book gives you a way to find the nth term of a geometric sequence, u_n:

$$u_n = ar^{n-1}$$

As usual, the formula book fails to define its terms: n is the number of the term you're looking for, a is the first term, and r is the common ratio.

For example, if you know that the first term is 3, the common ratio is 0.7, and you want the 11th term, you'd work the problem out as follows:

$$u_{11} = 3 \times 0.7^{10} \approx 0.0847$$

You can rearrange this formula to find a, r or n, given u_n and the other two. For example, if you know the sixth term of a geometric sequence is 10 and its common ratio is 0.5, you can work out $a = \dfrac{u_n}{r^{n-1}} = \dfrac{10}{0.5^5} = 320$.

Finding a sum

There are two sum formulas for the geometric series, which you use in different circumstances:

- ✔ **The sum to *n* terms:** Use this one when your geometric series finishes somewhere.

- ✔ **The sum to infinity:** Use this version when your geometric series goes on forever. You can use it only if the terms of the series get progressively closer to 0, which is to say that the common ratio is greater than −1 and less than 1.

The sum to n terms

Here's the formula for the sum to n terms of any geometric series:

$$S_n = \frac{a\left(1 - r^n\right)}{1 - r}$$

As usual, S_n is the sum of the first n terms, a is the first term, and r is the common ratio.

You use the formula in exactly the way you'd expect: if you're given a, r and n, you simply plug in the numbers and see what comes out. To work out the sum of the first ten terms of the geometric series beginning $4 + 6 + 9 + 13.5 + \ldots$, you would do the following:

1. **Write down *a*.**

   $a = 4$

2. **Work out *r* by dividing any term by the one before it.**

   $r = 6 \div 4 = 1.5$

3. **Plug the numbers into the formula.**

   $$S_{10} = \frac{4\left(1 - 1.5^{10}\right)}{1 - 1.5} \approx 453.32$$

You can also rearrange the formula, if needed, to work out any one of the missing variables given the others – although if you're working out r, it would need to be something fairly obvious for it to be a fair question!

The sum to infinity

If the terms in a geometric series are getting progressively smaller in magnitude (that is, each term is closer to 0 than the last one was), it turns out that their sum converges to a specific answer. It never quite reaches the specific answer, but it gets as close as you could ever want.

You've seen this phenomenon before: for example, you can think of 0.3333333 . . . as a geometric series if you write it like this:

$$\frac{3}{10} + \frac{3}{100} + \frac{3}{1,000} + \dots$$

Here, a is $\frac{3}{10}$, r is $\frac{1}{10}$, and after n terms of the series, the sum is

$$S_n = \frac{a(1-r^n)}{1-r}$$

$$= \frac{\frac{3}{10}\left(1-\left(\frac{1}{10}\right)^n\right)}{1-\frac{1}{10}}$$

Multiplying the top and bottom by 10 and then dividing both by 3 gives you

$$S_n = \frac{1-\left(\frac{1}{10}\right)^n}{3}$$

As n gets larger, $\left(\frac{1}{10}\right)^n$ gets smaller and smaller and smaller, and the sum gets as close as you could possibly want to $\frac{1}{3}$.

This limit is known as the *sum to infinity* of the series, and it has its own equation in the formula book:

$$S_\infty = \frac{a}{1-r}, \quad |r| < 1$$

As usual, the inequality is just there to tell you when the formula applies: it's valid only when the common ratio is between –1 and 1. Otherwise, the terms of the series aren't getting any closer to 0, and the sum of the series never settles down to a single value.

For example, if your first term is 2 and your common ratio is –3, the sum to one term is 2, the sum to two terms is $2+(-6)=-4$, and the sum to three terms is $2+(-6)+18=14$; these partial sums get progressively further from 0.

By contrast, if the first term is 2 and your common ratio is $-\frac{1}{3}$, the partial sums are $2, \frac{4}{3}, \frac{14}{9}, \frac{40}{27}, \frac{122}{81}, \dots$, which get closer and closer to $\frac{3}{2}$ the further you go.

There's a lot more than 'the terms get smaller' involved in whether a *general* series converges to a value. For example, the *harmonic series* $\frac{1}{1} + \frac{1}{2} + \frac{1}{3} + \frac{1}{4} + \dots$ has terms that get progressively smaller, but the total gets as big as you like (eventually – it diverges extremely slowly). However, geometric series with an appropriately small r always converge.

Finding parameters

It's going to sound simple when I say it: finding the parameters of a geometric series or sequence, given either some terms or the sum, is as simple as writing down equations for what you know and solving them. Naturally, the devil is in the detail: you have to write down the right equations, and you have to know how to solve them.

For example, if you're told the second and fifth terms of a geometric sequence are 8 and 27, you can work out the parameters a and r like this:

1. **Write down the information you have.**

   $u_2 = 8$ and $u_5 = 27$.

2. **Apply the relevant formulas.**

   The formula for the nth term of a sequence is $u_n = ar^{n-1}$, where a is the first term and r is the common ratio; therefore, the information about the second term gives you $ar = 8$, and the details about the fifth term give you $ar^4 = 27$.

3. **Solve the resulting simultaneous equations.**

   If you say $a = \frac{8}{r}$, then $\left(\frac{8}{r}\right)r^4 = 27$, so $r^3 = \frac{27}{8}$. That gives you $r = \frac{3}{2}$; substituting back into either equation gives you $a = \frac{16}{3}$.

You may be given details about the sum to a certain number of terms rather than just the terms themselves. For example, you might be asked to find possible values for both a and r if the second term of a geometric series is 6 and its sum to infinity is 49. Here's what you'd do:

1. **Write down the information you have.**

   $u_2 = 6$ and $S_\infty = 49$.

2. **Apply the relevant formulas.**

   $ar = 6$ and $\frac{a}{1-r} = 49$.

3. Solve the resulting simultaneous equations.

From the second equation, $a = 49(1-r)$, so the first equation becomes $49(1-r)r = 6$. Expanding and rearranging gives you $49r^2 - 49r + 6 = 0$, which factorises as $(7r-1)(7r-6) = 0$; therefore, $r = \frac{1}{7}$ or $r = \frac{6}{7}$. In the first case, a is 42; in the second, a is 7.

4. Check your answers are valid.

Both rs need to be between -1 and 1 for the infinite sum to exist. In this case, they both are. Also, watch out for conditions in the question that say something like 'r is positive'; here, there are no restrictions.

You're within your rights to use the quadratic formula on that quadratic rather than factorise it. However, factorising is more fun!

Awful questions

The examiners have scope, especially with geometric series, to ask awful questions. I'm not talking about the ones where they ask you to read a big block of text – after all, you're studying for A level, and it's reasonable to expect you to be able to read and pull information out of it without grumbling.

I mean the ones where they start throwing extra constants around, or the ones that ask you how far you need to go through the series before the sum exceeds a particular value. Luckily, you can solve the awful question with a little work. Let me show you how.

Finding parameters with unknown constants

A typical bloodbath Core 2 question says something like this:

> The first three terms of a geometric series are $4 + k$, k and $3k - 10$, where k is a positive constant. Find the value of k and the sum to infinity of the geometric series.

Well, then. How hard can it be?

1. Write down what you know.

$u_1 = 4 + k$, $u_2 = k$ and $u_3 = 3k - 10$.

2. Apply the appropriate formulas.

$a = 4 + k$, $ar = k$, and $ar^2 = 3k - 10$.

3. Solve the equations.

Here, you have simultaneous equations with three unknowns. But luckily, you're given a directly and can substitute that into the others: $(4+k)r = k$, so $r = \dfrac{k}{4+k}$. Substituting that into the third equation gives you

$$(4+k)\left(\frac{k}{4+k}\right)^2 = 3k - 10$$

which looks nastier than it is. One of the $(4+k)$s cancels out nicely; then you can rewrite the equation as

$$k^2 = (3k - 10)(4 + k)$$

Then expand and rearrange:

$$0 = 2k^2 + 2k - 40$$
$$0 = 2(k - 4)(k + 5)$$

That means $k = 4$ or $k = -5$.

4. Check if your answers make sense.

Here, $k = -5$ is an invalid answer because you're told k is a *positive* constant. That means k is 4.

5. Do the last bit of the question!

You can work out that a is 8 and r is $\dfrac{1}{2}$, meaning the sum to infinity is

$$\frac{a}{1-r} = \frac{8}{1 - \dfrac{1}{2}} = 16$$

First sum greater than S

A typical Core 2 question starts you off nice and gently – maybe giving you a and r and asking you to find a certain term or a sum to a certain number of terms. Then, casually, it'll drop in a bombshell like 'What is the smallest integer n such that the sum to n terms exceeds [some number]?'

Let's say $a = 3$ and $r = 2$, and you want to know the smallest n such that the sum to n terms is greater than 1,000. Here's what to do:

1. Write down what you know.

$$S_n > 1,000$$

2. Apply the appropriate formulas.

$$\frac{a\left(1-r^{n}\right)}{1-r} > 1,000$$

$$\frac{3\left(1-2^{n}\right)}{1-2} > 1,000$$

3. Solve the inequality.

$-3\left(1-2^{n}\right) > 1,000$, so $3 \times 2^{n} > 1,003$ and $2^{n} > \frac{1,003}{3}$. Take logarithms of both sides: $n\log(2) > \log\left(\frac{1,003}{3}\right)$, so $n > \log\left(\frac{1,003}{3}\right) \div \log(2) \approx 8.38$.

4. Give a sensible answer.

There's no such thing as an 8.38th term, and n is bigger than 8.38, so the smallest possible n is 9.

Be super careful with minus signs when you're dealing with inequalities, especially with logarithmic ones. If you ever need to divide or multiply by a negative number, remember to swap the direction of the inequality. Also remember that if $x < 1$, then $\log(x)$ is negative!

Proving the Sum Formulas

It's not common, but it's entirely possible for the examiners to ask you to prove the formula for the sum of the arithmetic series (usually in Core 1; Core 2 if you're doing OCR) or the sum of the geometric series (in Core 2, whichever board you're with).

The proofs themselves aren't all that difficult – it's just a case of remembering what to do. For each proof, there's a *nugget*, a key idea that you need to use, and everything else follows from that.

You probably want to memorise these proofs – or at least make sure you know the nuggets well enough to reproduce them.

Proving the arithmetic series sum

There's a story about the young Carl Friedrich Gauss adding up all the numbers from 1 to 100 while his teacher nursed a hangover. I won't tell it here, as it's in every other A level maths book ever written. However, his method and the method for proving that the sum of the arithmetic series is $\frac{1}{2}n(a+L)$ are pretty much identical.

The nugget is that if you write the series forwards and backwards, you get n pairs of numbers that all add up to the same thing. Here's how you prove it:

1. **Write down the sum to n terms of an arithmetic series explicitly.**

$$S = a + (a+d) + (a+2d) + \ldots + L$$

Leave plenty of space between the terms so you can align the next step nicely.

2. **Write down the same sum the other way around.**

$$S = L + (L-d) + (L-2d) + \ldots + a$$

Write each term under its 'partner' in the line above.

3. **Add the two together!**

You get

$$2S = (a+L) + (a+L) + \ldots + (a+L)_.$$

Lookit! They're all $(a+L)$! All n of them, in fact! So you could write them as $n(a+L)$.

4. **Rearrange.**

If $2S = n(a+L)$, then $S = \frac{1}{2}n(a+L)$, as required.

You can rewrite the last term using $L = a + (n-1)d$ if you need to put the sum in the other form, $S = \frac{1}{2}n(2a + (n-1)d)$.

Proving the geometric series sum

The nugget for the geometric series proof is to write out the sum and multiply it by r. Here's the proof, step by step:

1. **Write down the sum to n terms explicitly.**

$$S_n = a + ar + ar^2 + ar^3 + \ldots + ar^{n-1}$$

2. **Write down the same thing multiplied by r.**

$$rS_n = ar + ar^2 + ar^3 + \ldots + ar^{n-1} + ar^n$$

It's a good idea to line up your ars and so on so it's clear the middle terms will vanish when you subtract in the next step.

3. Take the second equation from the first.

You get $S_n - rS_n = a - ar^n$, because everything else vanishes!

4. Factorise both sides.

$$S_n(1-r) = a(1-r^n)$$

5. Divide by $(1-r)$, and you're done.

$$S_n = \frac{a(1-r^n)}{1-r}$$

Breaking Down the Binomial Expansion

If someone asked you to work out $(x-2)^2$, I trust you'd immediately expand $(x-2)(x-2)$ and get $x^2 - 4x + 4$. If she asked for $(x+5)^3$, you'd probably grumble and do it the same way. But if someone with an evil glint in her eyes asked you to work out $(3x+7)^{13}$, and poking her in her evilly glinting eyes wasn't an option, I hope you'd say, 'There has to be an easier way than multiplying out 13 horrible sets of brackets.'

You'd be absolutely right: it's something called the *binomial expansion*, which you get the general idea of in Core 2, and then you deal with nastier powers in Core 4.

Positive-integer powers

In Core 2, you only need to deal with positive-integer powers – that is, natural numbers. You'll be asked to expand things like $(2x+3)^5$ and $\left(1+\frac{x}{10}\right)^{10}$ but not $(2x+3)^{-5}$ or $(1+10x)^{1/10}$, which could conceivably come up in Core 4.

You're given a very helpful formula in the book:

$$(a+b)^n = a^n + \binom{n}{1}a^{n-1}b + \binom{n}{2}a^{n-2}b^2 + \ldots + \binom{n}{r}a^{n-r}b^r + \ldots + b^n$$

This is a bit of a mess, but I can tidy it up for you. The $\binom{n}{r}$ notation isn't anything to do with vectors; instead, it's to do with *combinations*. The '...' in the middle means you follow the same pattern as far as you possibly can, all the way up to $\binom{n}{n-1}a^1 b^{n-1}$, followed by the final term, b^n. When you expand an expression of the form $(a+b)^n$, you can confidently expect to get $n+1$ terms in your expansion.

You're given a formula for combinations, too: $\binom{n}{r} = \dfrac{n!}{r!(n-r)!}$, where $x!$ denotes the factorial, the product of all the natural numbers from 1 to x. For example, $4! = 4 \times 3 \times 2 \times 1 = 24$. However, this isn't critical knowledge for the binomial expansion: your calculator has all the buttons you need – and there's a way of doing it without troubling your calculator at all.

Using your calculator

There's another common way to write $\binom{n}{r}$: you can also write it as ^nC_r, and your calculator has a button for that! To work out $\binom{5}{3}$, you would type 5, then press SHIFT followed by '÷' (you should see a 'C' on your screen) and then 3. Pressing '=' gives you the correct answer of 10.

This method works only for positive-integer values of n and r – that is to say, the kinds of questions you'll come across in Core 2. See the upcoming 'Horrible powers' section for guidance on what to do if your powers are, um, horrible.

The 'nCr' button gives you the $(r-1)$th entry in the $(n-1)$th row of Pascal's triangle, which you may have seen and/or used; it's not a terribly practical method compared to others, so I don't show it here. As for what it actually *means*, it's the number of ways you can pick r objects out of n distinct possibilities, when you don't care about the order. For example, the number of possible five-card poker hands is $^{52}C_5$.

A sideways version of Pascal's triangle

There's a neat way to generate the $\binom{n}{r}$s without needing a calculator or the rows in Pascal's triangle. Here's how to do it for any n you like – let's say 7.

1. **Write down 1.**

   This is $\binom{n}{0}$ for any n you happen to pick.

2. **Multiply your answer by n and divide by 1; write down the answer.**

   You get n, which is 7 in this case. Stick with it, and you'll see why I wrote it that way.

3. **Multiply this answer by 1 less than you multiplied by in Step 2, and divide by 1 more than you divided by; write down the answer.**

   In this case, you multiply the 7 by 6 and divide by 2 to get 21.

4. **Repeat Step 3 as often as needed, each time multiplying by 1 less and dividing by 1 more.**

   The next step is to multiply the 21 by 5 and divide by 3 to get 35; then find $35 \times 4 \div 4$, and so on. The remaining numbers are 21, 7 and 1.

The eight numbers you generate are 1, 7, 21, 35, 35, 21, 7 and 1. You might notice the symmetry there: that's not a coincidence. In fact, once you've got halfway through your list, you can automatically write down the reverse of the first half to get the second.

Putting it together

Now, to find the expansion of something like $(2x - 3)^5$, here's what I would do:

1. **Draw out a table like the one in Figure 8-1. It will need four rows to write in and one more column than the power you're working with.**

   This table will have six columns.

2. **Call the first thing in the bracket a, the second thing b and the power n.**

   In this case, $a = 2x$, $b = -3$ and $n = 5$.

3. **Work out all the $\binom{n}{r}$s you need for the expansion, using any method from the preceding sections, and write them down in the first row of the table.**

   Your numbers here are 1, 5, 10, 10, 5 and 1.

4. **Work out a^n and write it at the start of the second row.**

   In each following column, write down a raised to one less power – here, you'd have $32x^5, 16x^4, 8x^3, 4x^2, 2x$ and 1, which is a^0. (An alternative method is to divide by a each time.)

5. **Write 1 at the start of the third row, and repeatedly multiply by b to get the following columns.**

   Here, you get 1 (which is b^0), –3, 9, –27, 81 and –243.

6. **Multiply each column together and add up the results.**

   You get $32x^5 - 240x^4 + 720x^3 - 1,080x^2 + 810x - 243$.

1	5	10	10	5	1
$32x^5$	$16x^4$	$8x^3$	$4x^2$	$2x$	1
1	–3	9	–27	81	–243
$32x^5$	$-240x^4$	$720x^3$	$-1,080x^2$	$810x$	–243

Figure 8-1: A table for evaluating a binomial expansion.

© John Wiley & Sons, Inc.

It's worth checking your expression works. Put a value for x into your expression and into the original to make sure you get the same answer: for $x = 1$, for example, you find both $(2x - 3)^5$ and the expanded version, $32x^5 - 240x^4 + 720x^3 - 1{,}080x^2 + 810x - 243$, work out to -1.

You may be asked about the *coefficient* of a given term, which is simply the number involved. In the example you just did, the coefficient of x^3 is 720.

Horrible powers

In Core 4, you extend the binomial series beyond those nice, safe, positive whole-number powers into the scary world of the fraction, the negative integer and – *shudder* – the negative fraction. Luckily, all three types work by the same method.

One thing to watch out for with these horrible powers is that the expansion is not defined for all values of x. If you have $(a + bx)^n$, where n is not a positive integer, then the binomial expansion is valid only when $|bx| < |a|$ – loosely speaking, the bit with the x in it can't be bigger (in magnitude) than the bit without the x in it.

In the formula book, you're given this *radius of convergence* in a slightly different form: if you've manipulated the expression into a form that looks like $k(1 + x)^n$, then $|x| < 1$. This works out to be the same thing.

The other thing to note about these nasty-powered binomial expansions is that, unlike their positive-integer counterparts, they never stop – you get an infinite series. Luckily, you're usually asked only for the first few terms – or else it'd be a very long and tedious exam. In most cases where the expansion is used, the terms tend to get very small very quickly, so you can get hold of a decent approximation with only a few terms.

The book way with $(1 + x)^n$

Here's what one formula in the book says:

$$(1+x)^n = 1 + nx + \frac{n(n-1)}{1 \times 2}x^2 + \ldots + \frac{n(n-1) \times \cdots \times (n-r+1)}{1 \times 2 \times \cdots \times r}x^2 + \ldots, \quad \left(|x| < 1, n \in \mathbb{R}\right)$$

Doesn't that look fun? In a sense, it's not *that* bad, at least if you're working with $(1 + x)$ in the bracket. To work out the binomial expansion of $\sqrt{1 + x}$ up to the term in x^3, you simply turn it into $(1 + x)^{1/2}$ and replace all the ns in the book equation with $\frac{1}{2}$:

$$(1+x)^{1/2} \approx 1 + \frac{1}{2}x + \frac{\frac{1}{2}\left(\frac{-1}{2}\right)}{2}x^2 + \frac{\left(\frac{1}{2}\right)\left(\frac{-1}{2}\right)\left(\frac{-3}{2}\right)}{6}x^3, \quad |x| < 1$$

You need to say that $|x| < 1$ because the series doesn't converge outside that interval; you don't need to say $n \in \mathbb{R}$, because you've specified n.

You can simplify the series, of course:

$$(1+x)^{1/2} \approx 1 + \frac{1}{2}x - \frac{1}{8}x^2 + \frac{1}{16}x^3, \quad |x| < 1$$

However, that step is much less simple than I've made it look. It's incredibly easy to lose a minus sign or squash your work in together so tightly it's impossible to read back.

The book way with $(a+bx)^n$

It gets worse if you have something like $\dfrac{1}{\sqrt{9-8x}}$, which is something that has come up in past papers. You have to jump through hoops to do it the book way, which involves manipulating your expression so it looks like $k(1+y)^n$. Here's what you'd do:

1. First, get it all into a 'nice' index form.

$$(9-8x)^{-1/2}$$

2. Now deal with the a.

You're aiming to get the first number in the brackets to be 1:

$$(9-8x)^{-1/2} = \left(9\left(1-\frac{8}{9}x\right)\right)^{-1/2}$$
$$= 9^{-1/2}\left(1-\frac{8}{9}x\right)^{-1/2}$$
$$= \frac{1}{3}\left(1-\frac{8}{9}x\right)^{-1/2}$$

Lovely!

3. Apply the formula to the bracket.

Ignoring the $\frac{1}{3}$ for the moment, you now substitute $-\frac{8}{9}x$ for x in the formula and get

$$1 + \left(-\frac{1}{2}\right)\left(-\frac{8}{9}x\right) + \frac{\left(-\frac{1}{2}\right)\left(-\frac{3}{2}\right)}{2}\left(-\frac{8}{9}x\right)^2 + \frac{\left(-\frac{1}{2}\right)\left(-\frac{3}{2}\right)\left(-\frac{5}{2}\right)}{6}\left(-\frac{8}{9}x\right)^3 + \ldots$$

4. Giggle hysterically. Then regather your composure.

5. Simplify the monstrosity.

I get

$$1+\left(-\frac{1}{2}\right)\left(-\frac{8}{9}\right)x+\left(\frac{3}{8}\right)\left(\frac{64}{81}\right)x^2+\left(-\frac{5}{16}\right)\left(-\frac{512}{729}\right)x^3+\dots$$

$$=1+\frac{4}{9}x+\frac{8}{27}x^2+\frac{160}{729}x^3+\dots$$

6. Oh, yes, don't forget to multiply it all by $\frac{1}{3}$ at the end.

Your final answer is $\frac{1}{3}+\frac{4}{27}x+\frac{8}{81}x^2+\frac{160}{2,187}x^3+\dots$, for $\left|-\frac{8}{9}x\right|<1$, or (better) $|x|<\frac{9}{8}$.

I'm a proper, card-carrying mathematician with two degrees, an equation named after me and a fearsome reputation for mental arithmetic – and *I* can't do the binomial expansion the book way the first time without making mistakes. In that example, I initially forgot to square and cube the $\frac{8}{9}x$ parts, *and* I lost a couple of minus signs (although I got lucky: they cancelled out). There are so many places to slip up! I'm happy to inform you that there's a nicer, more reliable way.

A much, much nicer way

This way of doing binomial expansions isn't in the formula book. I don't know why it's not generally taught. It should be. It's based on the sideways version of Pascal's triangle from earlier in this chapter, so make sure you understand that; the only small difference is that the number you start with isn't a positive integer. For the example in the preceding section, $(9-8x)^{-1/2}$, you would need to work out the multipliers for $n=-\frac{1}{2}$.

You start with 1 and then $-\frac{1}{2}$. You multiply this by 1 less than $-\frac{1}{2}$, which is $\left(-\frac{3}{2}\right)$ and divide by 1 more than 1 (2) to get $-\frac{1}{2}\times\left(-\frac{3}{2}\right)\div2=\frac{3}{8}$. Multiplying this by 1 less than $-\frac{3}{2}$ and dividing by 1 more than 2 gives you $\frac{3}{8}\times\left(-\frac{5}{2}\right)\div3=\frac{5}{16}$, which is as far as you need to go for this problem.

Here's what you do:

1. Draw out a table with four rows and as many columns as you need terms.

Here, you need four terms, so four columns is plenty.

2. Call the first thing in the bracket a, the second thing b and the power n.

In this case, $a=9$, $b=-8x$ and $n=-\frac{1}{2}$.

3. **Write down your multipliers in the first row of the table, like the one in Figure 8-2.**

   Your top row should read $1, -\dfrac{1}{2}, \dfrac{3}{8}, -\dfrac{5}{16}$.

4. **Work out a^n and write it at the start of the second row, under the first 1.**

   In each following column, write down a raised to one less power – here, you have $\dfrac{1}{3}, \dfrac{1}{27}, \dfrac{1}{243}$ and $\dfrac{1}{2,187}$. (An alternative method to generating these numbers is to divide by a each time.)

5. **Write 1 at the start of the third row, and repeatedly multiply by b to get the following columns.**

   Here, you get $1, -8x, 64x^2$ and $-512x^3$.

6. **Multiply each column together and add the results together.**

   You get $\dfrac{1}{3} + \dfrac{4}{27}x + \dfrac{8}{81}x^2 + \dfrac{160}{2,187}x^3 + \ldots$, the same answer as the other way but with a tiny fraction of the work and confusion.

Figure 8-2:
Working
table for
binomial
expansions.

1	$-\dfrac{1}{2}$	$\dfrac{3}{8}$	$-\dfrac{5}{16}$
$\dfrac{1}{3}$	$\dfrac{1}{27}$	$\dfrac{1}{243}$	$\dfrac{1}{2,187}$
1	$-8x$	$64x^2$	$-512x^3$
$\dfrac{1}{3}$	$\dfrac{4}{27}x$	$\dfrac{8}{81}x^2$	$\dfrac{160}{2,187}x^3$

© John Wiley & Sons, Inc.

Estimating with the Binomial Expansion

The main use of the binomial expansion – before we had calculators that could do the sums for us – was to work out estimates of awkward calculations such as 2.1^6. In this case, you would work out $(2+x)^6$ and let $x = 0.1$.

Here's how you'd do that, in a more precise recipe form:

1. **Work out the binomial expansion you're asked for.**

   This is usually part a) of the question, and generally only a few terms will be needed. Here, $(2+x)^6 = 64 + 192x + 240x^2 + 160x^3 + \ldots$.

2. Work out what x needs to be to make your original bracket the same as (or an easy multiple of) the thing you're trying to work out.

In this example, $2 + x = 2.1$, so $x = 0.1$.

3. Substitute your value for x into the expansion.

You get $64 + 19.2 + 2.4 + 0.16 = 85.76$.

4. Check your answer is reasonable on the calculator.

$2.1^6 = 85.77$ (to two decimal places), so you have a pretty good approximation (if not quite exactly right).

A variation on this method involves using the binomial expansion to find a root or a multiple of it. For example, you might be asked first to show that the square root of 0.96 is $\frac{2}{5}\sqrt{6}$ (which it is) and then to use the first four terms of the binomial expansion of $(1 - 4x)^{1/2}$ to find an estimate of $\sqrt{6}$. Here's the drill:

1. Do the first thing they say.

To show $\sqrt{0.96} = \frac{2}{5}\sqrt{6}$, work with fractions:

$$\sqrt{\frac{96}{100}} = \sqrt{\frac{4^2 \times 6}{10^2}} = \frac{2}{5}\sqrt{6}$$

2. Now work out the binomial expansion they suggest.

$$(1 - 4x)^{1/2} \approx 1 - 2x - 2x^2 - 4x^3$$

3. Use the binomial expansion to work out the square root of the number they mention.

Here, you're looking for $\sqrt{0.96}$; you can work this out if you let $1 - 4x = 0.96$. You get $x = 0.01$, which you can substitute into the expansion, so

$$\sqrt{0.96} \approx 1 - 0.02 - 0.0002 - 0.000004 = 0.979796$$

4. Use this to work out the number they're after.

Because $\sqrt{0.96} = \frac{2}{5}\sqrt{6}$, you can say $\sqrt{6}$ is $\frac{5}{2}$ the size of $\sqrt{0.96}$. Multiply your estimate by $\frac{5}{2}$ to get 2.449490, which is correct to six decimal places.

If they ask for some number of decimal places, give them that many decimal places, even if it involves padding the end of the number with zeros, like I have here.

Chapter 9

Fiddling About with Functions

· ·

In This Chapter

▶ Understanding functions and their notation

▶ Composing functions and finding their inverses

▶ Doing iterations, root-bounding and numerical integration

· ·

*T*he function is one of the most important ideas in mathematics. Pretty much anything you do in A level maths involves a function at some level – even if it's a hidden function like 'add' or 'to the power of'. So what is it?

A *function* (if you ask a textbook) is a mapping from one set of things (the *domain*) to another (the *codomain*), such that each element in the first set is mapped to exactly one thing in the other. You're quite within your rights to read that and then want to throw your textbook away; that's just how mathematicians talk.

I prefer to think of a function in terms of a certain kind of quiz. The set of possible questions is the *domain* – what you're allowed to ask the function. The set of possible answers is the *codomain*, what you could conceivably get out. And the function is simply the link between each question and its unique correct answer.

For example, a function might map a year (the domain) to a football team (the codomain), giving the winner of the Scottish FA Cup in that year. Applying the function to the domain element '1961' would result in 'Dunfermline Athletic'.

A more mathematical function might be something like $f(x) = x^2 - x$ for $x \in \mathbb{R}$. In the next section, I break down what that means – including the funny symbols – and explain how the function works in detail. I then go on to more advanced functions questions, which involve *composing* functions (applying another function to the result of the first) and *inverting* functions (undoing them).

There are also more evil questions to be answered, and – only tangentially related to functions – numerical methods for finding solutions to equations.

Putting the 'Fun' in Functions

Defining a function requires three parts:

- ✔ **The name of the function:** Usually, these are things like f, g or – just for the sake of variety – h. They're not restricted to that, though: any letter can be a function, and things like sin, sec and $\tan^{-1}$ are all perfectly good names. Giving functions names means we have a useful shorthand for referring to a specific machine.

- ✔ **The mapping:** This is the bit that gets most of the attention: the recipe that describes what happens to your variable when you apply the function to it.

- ✔ **The domain:** This tells you what values you're allowed to put into the function.

In this section, I show you exactly how the notation for a function works and how you can figure out what's going on with domains and *ranges*, the possible set of results of a function. (I don't figure you need much help with the names part of things.)

Nailing down the notation

There are two common ways to see a function defined. They're very similar, and there's no real difference in meaning between the two. You can use them interchangeably.

The 'proper mathsy one' involves colons and arrows, and it looks like this:

$$f : x \to \frac{x^2+1}{x-1}, \quad x \in \mathbb{R}, \ x \neq 1$$

It's really the same thing as this:

$$f(x) = \frac{x^2+1}{x-1}, \quad x \in \mathbb{R}, \ x \neq 1$$

Both mean that f is a machine that takes in a value (call it x), squares it and adds 1, then divides it by 1 less than the original number. For example, $f(0) = -1$ and $f(2) = 5$.

The bit on the right – the *domain* – tells you that x is any real number except for 1 (the funny R means the set of real numbers – for now, you can understand it as 'any number you can, in principle, approximate as a decimal'). You have to disallow 1 in this function, or else some smart-aleck would try to divide by 0, and we can't be having that now, can we?

Dealing with the domain

For a function to be properly defined, it really ought to have a domain attached to it. The *domain* of a function is 'the set of values you're allowed to put in'. As well as the 'all real numbers except. . .' sort of domain (see the preceding section), there's also the 'inequality' sort of domain, which is probably more familiar. For example:

$$f(x) = \ln(x) + \ln(4-x), \quad 0 < x < 4$$

That bit after the comma means that all the *x*-values you're allowed to put into the function are between 0 and 4 (exclusive, in this case) – you can put in 2 or π or 0.123 but not -5, 7 or even 0.

The inequality may be open-ended, like this one:

$$f(x) = \sqrt{x-4} \quad x \geq 4$$

This simply means that any *x* you put into the function has to be at least 4. Values of 7, 4.5 and 4 are fine, but 0, 3 and -6 are no good. (In this case, it's because anything smaller than 4 will break the square root, but there's nothing to stop the domain from being $4 \leq x \leq 2\pi$ or whatever else you wanted – as long as the square root is defined.)

Most of the functions you're likely to see have a *natural* domain – anything you could put into the function without breaking maths. You won't see a question asking about a natural domain in so many words; instead, you might be asked what values need to be excluded from the domain, or you may be asked to suggest a domain. Here are some things you'll need to look out for:

- ✔ **Dividing by things:** You need to exclude anything that would make the bottom of a fraction 0.

- ✔ **Taking logarithms:** If you have a logarithm, you have to exclude anything that isn't strictly positive. (Zero is *not* allowed in a logarithm.)

- ✔ **Doing even roots (and fractional powers with an even denominator):** If you have a square root, a fourth root, or anything similar, you have to exclude negative numbers (but 0 *is* allowed).

For an example that brings all these things together, consider the function

$$f(x) = \frac{\sqrt{x^2 - 4}}{\ln(3x + 10)}$$

What's its natural domain?

1. **Look for any divisions and exclude anything that makes the bottom 0.**

    If $3x + 10 = 1$, then the bottom is 0 (because $\ln(1) = 0$). That means x can't be -3.

2. **Look for any logarithms and exclude anything that makes the argument 0 or smaller.**

    If $3x + 10 \leq 0$, you have a bad problem, so you need to exclude $x \leq -\frac{10}{3}$.

3. **Look for any even roots and exclude anything that makes them negative.**

    Here, you're looking for $x^2 - 4 < 0$, which happens for $-2 < x < 2$, so you need to exclude those values, too.

4. **The natural domain is everything else.**

    Here, the natural domain is $-\frac{10}{3} < x < -3$, $-3 < x \leq -2$ or $x \geq 2$. See Figure 9-1.

This is a contrived example to show how it works – you'll not see anything this complex in an exam.

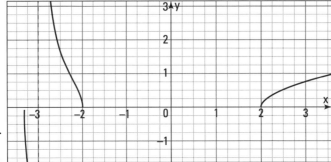

Figure 9-1:
Finding the natural domain of $y = \dfrac{\sqrt{x^2 - 4}}{\ln(3x + 10)}$.

© John Wiley & Sons, Inc.

Finding the range

The *range* of a function is the set of possible outputs you get when you enter the inputs. This is not quite the same thing as the codomain, which is everything you could *potentially* get out of the function; the range is only the values you can *actually* get out.

Finding the range of a function is a bit of an art form, but here's how I'd approach it:

1. **Find the output of the function at both ends of the domain.**

2. **If either end of the domain is unspecified (for example, $x > 0$ has no upper limit), see what happens when x is very large (or very small).**

3. **Find any vertical asymptotes (usually where the bottom of a fraction would be 0 or the argument of a logarithm would be 0) and see what the values of the function are if you change x to be very slightly larger or smaller.**

4. **Find the coordinates of any turning points in the domain.**

5. **Sketch the curve $y = f(x)$!**

6. **Find the lowest and highest points your graph goes to, and see whether there are any y-values between them that the graph doesn't have.**

   For example, the graph of $y = \dfrac{1}{x}$ has a horizontal asymptote at $y = 0$, and the graph of $y = \sec(x)$ never reaches a y-value strictly between -1 and 1.

Be very careful about whether a limit is actually reached or not; this determines whether you need a strict ($<$ or $>$) or inclusive ($\le$ or $\ge$) inequality. Some tips:

- ✔ If your extreme value occurs at either end of the domain, use the same kind of inequality as the domain (that is, if the domain uses a strict inequality, so should the range).

- ✔ If your extreme value occurs at a turning point, use an inclusive inequality (the curve reaches the value, so the value is included in the range).

- ✔ If your extreme value occurs as x goes to positive or negative infinity, use a strict inequality (the curve never quite reaches the asymptote).

You should give your answer as an inequality or set of inequalities (just like the domain) but involving $f(x)$ instead of just x: the range of $f(x) = x^2$ would be $f(x) \ge 0$.

Here are some further examples of finding the range:

- ✔ If $g(x) = \dfrac{1}{x}$, $x \ge 3$, the range would be $0 < g(x) \le \dfrac{1}{3}$, because the highest value $\left(\dfrac{1}{3}\right)$ occurs at the left end of the domain, which is included. There are no turning points, and the values get closer to 0 as x gets large, without ever quite reaching the axis.

✔ If $h(x) = e^x - xe$, $x > 0$, you'd draw a sketch like the one in Figure 9-2a. After finding the turning point (see Chapter 15) at $(1, 0)$, you'd be able to say that the range was $h(x) \geq 0$.

✔ If $j(x) = x^2 - 4x + 3$, $-5 \leq x \leq 5$, you'd draw a sketch like the one in Figure 9-2b. The vertex is at $(2, -1)$, and the ends of the domain give you $j(5) = 8$ and $j(-5) = 48$. The range is thus $-1 \leq j(x) \leq 48$.

(a)

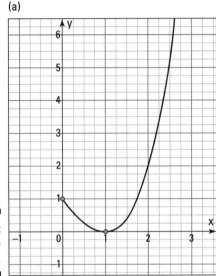

(b)

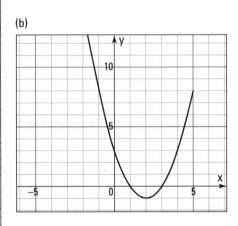

Figure 9-2: Finding the range.

Composing and Inverting Functions

The idea of a function is a very useful one – but it's made even more useful by the ideas of chaining functions together (*composing* functions to make *composite functions*) and running them backwards (*inverting* functions to get their *inverse functions*).

In this section, I take you through the kinds of things you can expect to see in Core 3 exam questions on these topics: how to combine functions in as clear a way as possible and how to find the 'undo' button on a function (if it exists!).

Composition: A chain of machines

To *compose* two functions, you apply one to the other. For example, if you have $f(x) = e^x$ and $g(x) = \cos(x)$, both defined for any real value of x, you can work out $f(g(x)) = e^{\cos(x)}$ and $g(f(x)) = \cos(e^x)$.

When composing functions, the order matters: $g(f(x))$ does not, generally, give the same answer as $f(g(x))$. (Try putting $x = 0$ into the preceding examples, and you'll see.)

There are an awful lot of brackets there. Because mathematicians are lazy, it's more usual to leave the outermost set of brackets out – instead of $f(g(x))$, you'd write $fg(x)$ or (rarely) $f \circ g(x)$. You may even see plain *fg*, which is the name of the function that results from the composition.

A normal Core 3 question might give you two functions and ask you to express their composition. For instance, if the examiners defined the following:

$$f : x \to 3x + \ln(5), \quad x \in \mathbb{R}$$
$$g : x \to e^{3x}, \quad x \in \mathbb{R}$$

they might ask you to show that $gf(x) = 125e^{9x}$. Here's how:

1. Write down $g(F)$ explicitly.

$$g(F) = e^{3F}$$

2. Replace every F with $f(x)$, explicitly.

$$gf(x) = e^{3(3x + \ln(5))}$$

3. Tidy up.

That's $e^{9x} \times e^{3\ln(5)} = 125e^{9x}$.

It's totally OK to use your calculator to work out $e^{3\ln(5)}$ in the exam, but if you do it while revising, I'll be over here rolling my eyes. You should be able to simplify that using your logarithm laws from Chapter 5 – it's good practice for when you have letters rather than numbers up top.

Finding the domain and range of composite functions is simpler in practice that it is in theory. In practice, the domain of *fg* is almost always the domain of *g*, which is the first machine you put values into. You calculate the range the same way as you would for any other function.

In *theory*, you need to make sure that the output you get from *g* is a valid input for *f*, but I've never seen an exam question that required this. In the unlikely event that you see one, you have to exclude from *g*'s domain any *x* such that $g(x)$ is not in *f*'s domain. You would be walked through the detail of such a question!

Inverses: Running the machines backwards

A function's *inverse* is the function that 'undoes' the original function – you're familiar with a few of these already: with appropriate restrictions that I'll get to in a minute, the inverse of $\tan(x)$, for example, is $\tan^{-1}(x)$; the inverse of e^x is $\ln(x)$; the inverse of x^2 is $\sqrt{x}$; and so on. The idea is that if you put a value into a function and then put the result into the function's inverse, you should get back to where you started. (It works both ways, incidentally: if you put the value into the inverse function and put the result of that into the original function, you'll also get your original value back.)

Translated into maths, the inverse function $f^{-1}(x)$ is the function such that $f^{-1}[f(x)] = x$ and $f[f^{-1}(x)] = x$, for all possible values of *x*.

In this section, I talk you through working out whether an inverse function exists and how to find the inverse if it does. I also tell you what to do about graphs of inverse functions, and I revisit domains and ranges.

Asking, 'Is there an inverse?'

Before I show you how to find an inverse, I have to show you how to tell whether an inverse function exists. For example, the function $f(x) = x^2$, $x \in \mathbb{R}$, does *not* have an inverse function.

You might expect the square root to be the inverse, but there's a subtle problem with that! Think about what happens with $x = -2$. Calculating $f(x)$ gives you $(-2)^2 = 4$; applying the square root function to 4 gives you 2. See the problem? The inverse doesn't take you back to your original value.

The function *f* is an example of a *many-to-one* function – several different inputs can give the same output. It can be turned into a *one-to-one* function, where each input gives a different answer, though: all you need to do is change the domain! The function $g(x) = x^2$, $x \geq 0$ *does* have an inverse function – every output of the function is associated with exactly one input.

A function can have an inverse only if it's *one-to-one*: no output value can be associated with more than one input, as illustrated in Figure 9-3.

(a)

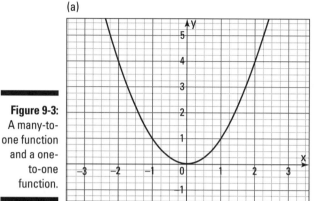

(b)

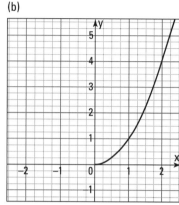

Figure 9-3:
A many-to-
one function
and a one-
to-one
function.

Finding an inverse

You can almost guarantee that a Core 3 paper will ask you to find the inverse of a function. A typical question will give you a function like this:

$$f : x \rightarrow \ln(9 - 3x), \quad x < 3$$

and ask you to find the inverse function. Here's what I suggest:

1. **The definition of the inverse function says that $f\left(f^{-1}(x)\right) = x$. Let $F = f^{-1}(x)$ and write down $f(F)$ explicitly.**

   $$f(F) = \ln(9 - 3F) = x$$

2. **Rearrange to solve for F.**

   Here, $9 - 3F = e^x$, so $F = \dfrac{9 - e^x}{3}$.

3. **This is the inverse function! Now write it in the proper format, and possibly tidy up.**

   $$f^{-1} : x \rightarrow 3 - \frac{1}{3}e^x$$

You could just as easily use the 'equals' function notation rather than the 'arrow' one and write $f^{-1}(x) = 3 - \frac{1}{3}e^x$. Or you could leave it all in a single fraction; unless you've been told to state the inverse function in a particular form, any correct answer (within reason) is fine.

There's one more 'special case' trick you can use in the right circumstance: if you know $ff(x) = x$, then f is its own inverse!

Looking at graphs of functions and their inverses

In Figure 9-4, I've drawn a few functions and their inverses: $y = e^x$ and $y = \ln(x)$ on the top; $y = x^2$ (for $x \geq 0$) and $y = \sqrt{x}$ in the centre; and $y = \cos(x)$ (for $0 \leq x \leq \pi$) and $y = \cos^{-1}(x)$ on the bottom. I've also drawn the line $y = x$ on all three. Notice anything?

The graphs of the original functions and their inverses are reflections in the line $y = x$; in effect, you're switching the x for the y and vice versa.

This comes in handy when you're asked to sketch the graph of an inverse function. The simplest steps I know are these:

1. **If necessary, draw some axes and sketch the original curve lightly.**

    Draw big, and label the important points (the ends of the domain, where the function crosses the axes, any turning points, any asymptotes – see the DATAS method in Chapter 10 for more on curve sketching).

2. **Draw the line $y = x$. More firmly, this time.**

3. **Sketch the inverse function.**

    Take every point whose coordinates you know in the original curve and swap them – $(0,1)$ becomes $(1,0)$ and so on. Plot these. If they don't make an obvious shape, find the coordinates of some more points on the original curve, swap the xs for the ys, and plot them until you *do* get an obvious shape.

4. **Join the dots, firmly, to make the graph of the inverse function.**

5. **As a check, turn your paper so the $y = x$ line is vertical; the curves of the original function and the inverse should be reflections of each other.**

Swapping domains and ranges

Knowing that the graph of the inverse function is simply a reflection in the line $y = x$ makes a particular kind of question easier to visualise: the kind that asks about domains and ranges of inverse functions.

Because reflecting in the line $y = x$ effectively swaps the axes, it swaps the domain and range:

✔ The domain of $f^{-1}(x)$ is the same as the range of $f(x)$.

✔ The domain of $f(x)$ is the same as the range of $f^{-1}(x)$.

A typical question will have you work out the range of the original function and ask you to write down the domain of the inverse function for one mark – 'write down' means you don't need to show any working; you just need to write down the range you previously worked out.

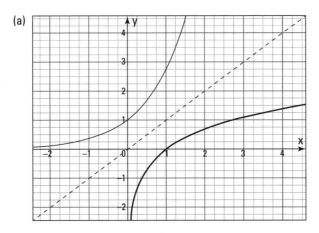

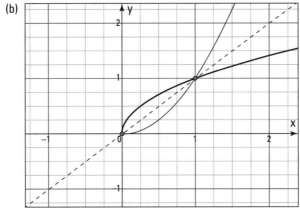

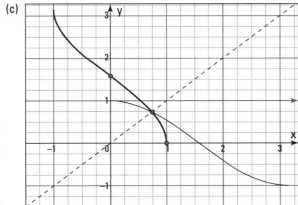

Figure 9-4:
Functions
and
inverses.

A few pages ago, I told you that many-to-one functions don't have an inverse. This is another reason why! If you swapped the domain and range of a many-to-one function, you'd get a 'function' that had two or more outputs for a single input, which is a big no-no.

y = x: Where functions meet their inverses

A particularly sneaky question that sometimes comes up in an exam is 'Where does $f(x) = f^{-1}(x)$?' – that is, where does the graph of the original function meet the graph of the inverse?

What you *shouldn't* do is try to work it out as it stands – in principle, you *could* write down the function and its inverse explicitly and solve the resulting mess, but it's more work than you need to do.

Look back at Figure 9-4 and spot where the functions and their inverses cross: it's always on the line $y = x$. Rather than solving $f(x) = f^{-1}(x)$, you can get away with solving $f(x) = x$ or even $f^{-1}(x) = x$, if that looks easier.

Making Sense of the Modulus

The *modulus* function doesn't really fit here, but it doesn't fit anywhere else in the book, either. This helpful function turns its argument positive. Its notation is a little unusual in that the modulus doesn't really look like a function; it looks like a vertical bracket, like this: $|x|$.

You can take $|x|$ to mean 'if x is positive, leave it alone; if it's negative, turn it positive.' For example, to work out $|5|$, you note that 5 is already positive and output the value 5. To work out $|-7|$, you say, 'Because -7 is negative, I need to turn it positive, so the output must be 7.'

There's a nice way to write this using *cases*:

$$|x| = \begin{cases} x, & x \geq 0 \\ -x, & x < 0 \end{cases}$$

This means exactly what I just said: when x is positive (or 0), the function outputs x; when it's negative, the function outputs $-x$, which turns the input into a positive number.

You can think of the modulus function as 'how far away from the origin (0) this number is'.

Most modulus-related questions rely on drawing a sketch, which I cover in Chapter 10. However, you may need to do some algebra with these problems, too.

Matching moduli

A typical algebra question involving moduli asks you to solve an equation involving at least one modulus – for example, $|3x-12|=6$. You can approach this in several ways, but the method that seems to work best for most people is the *plus-minus* method.

The main advantage of this method is that it doesn't involve very much thinking; the disadvantage is that you have to be careful about checking your answers really are answers. Here's how it goes:

1. **Try solving the equation without the modulus signs.**

   Here, you solve $3x-12=6$ to get $x=6$.

2. **Pick a modulus bracket, replace it with regular brackets preceded by the opposite sign, and solve the resulting equation.**

   In this example, replace $|3x-12|$ with $-(3x-12)$ to get $-(3x-12)=6$; the solution to this equation is $x=2$.

3. **If there are two or more modulus brackets, do the same thing for each of them.**

   Each of the modulus brackets could be positive or negative. If you're finicky about it, only certain combinations make sense, but it's generally less hassle just to try all the combinations and check rather than work out ahead of time which ones are sensible.

4. **Check that all your answers satisfy the original equation.**

   In some situations, the plus-minus method gives spurious answers, so you need to make sure they really work. Here, substituting $x=6$ into the original left-hand side gives you $|6|$, which is indeed 6; substituting in $x=2$ gives you $|-6|$, which is also 6, so both answers are valid.

   Sketching what you're looking at is an excellent habit to have in general, and it's a great check that the answers you're getting make sense. You find a section on drawing graphs involving the modulus function in Chapter 10.

Handling inequalities with a modulus

If you have an inequality involving a modulus sign, you need to do a bit more work, but the basics are similar. Given the problem $|x-4|>|3+2x|+1$, here's what you do:

1. **Find where the expressions on either side of the inequality are equal (use the plus-minus technique from the preceding section).**

   Your answers are $x=0$ and $x=-6$.

2. Solve the inequalities (use either method from Chapter 6).

If you use the sketch method, you can clearly see that the graph of the right-hand side of the given inequality has a lower *y*-value between the solutions (see Figure 9-5); therefore, the set of solutions is $-6 < x < 0$. (See Chapter 10 for details on sketching modulus graphs.)

Alternatively, using the check-between method, you find that for *x* below -6, the right-hand side of the inequality is greater; between 0 and -6, the left-hand side is greater (as you want); and above 0, the right-hand side is greater again.

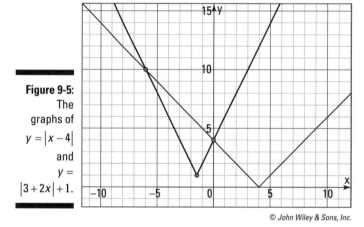

Figure 9-5:
The graphs of
$y = |x - 4|$
and
$y = |3 + 2x| + 1$.

Evil Functions Questions

You can do loads of things with functions in maths: you can reach all sorts of deep results about the infinitely large, the infinitesimally small and the anywhere-in-between. But instead, exam boards just ask you nasty questions.

In this section, I take you through some of the more common nasty questions: solving something involving composed functions, what to do when an inverse is involved, and – for the benefit of students doing the one board that covers them – even and odd functions.

Solving for composed functions

Working backwards with functions is something you certainly have the skills for, but it somehow always seems tricky for students, so I'll run you through an example to be on the safe side. Suppose you've been told that $f(x) = \dfrac{1}{x+2}$ (for all real numbers except -2) and that $g(x) = 4x^2 - 3$ (for all real numbers), and you want to solve for where $fg(x) = \frac{1}{3}$. Here's how I'd do it:

1. **Compose the functions like you've been doing all along.**

   $f(G) = \frac{1}{G+2}$, so $fg(x) = \frac{1}{4x^2 - 1}$. (See the earlier section 'Composition: A chain of machines' for details.)

2. **Solve for where the composite function gives you the required value.**

   $\frac{1}{4x^2 - 1} = \frac{1}{3}$, so $4x^2 - 1 = 3$ and $x = \pm 1$.

3. **Check that each answer for x is in the domain of g. Also check that when you work out $g(x)$ for each x, the result is in the domain of f.**

   In this case, both answers are in the domain of g; $g(x) = 1$ for both x-values, and 1 is in the domain of f.

It's important to check that your answers go smoothly through the functions, or else you can end up with incorrect answers.

Combining inverses and functions

One type of question has an especially elegant way to solve it (although there's a more brute-force method as well). Suppose you're given that $f(x) = 4e^{2x+1}$ and $g(x) = 3x + 4$, both defined for all x, and the question asks you to solve $f^{-1}g(x) = \ln(2)$.

The obvious way to do it is to work out the inverse function of f, apply it to $g(x)$ and solve for x. That's perfectly doable, but there's a simpler way. Here it is:

1. **Apply $f(x)$ to both sides.**

   $$f\left(f^{-1}g(x)\right) = f\left(\ln(2)\right)$$

   The inverse function on the left cancels out to leave you with

   $$g(x) = f\left(\ln(2)\right)$$

2. **Work out the right-hand side.**

   $$g(x) = f\left(\ln(2)\right)$$
   $$= 4e^{2\ln(2)+1}$$
   $$= 4\left(e^{\ln(4)} \times e^1\right)$$
   $$= 16e$$

3. Now replace $g(x)$ with its definition and solve for x.

$3x + 4 = 16e$, so $x = \dfrac{16e - 4}{3}$.

You can use this trick whenever the function side starts with an inverse function!

Even and odd functions

I'll give you an example of an even function: $f(x) = x^2$. And another: $g(x) = x^{10}$. An odd function? How about $h(x) = x^3$ or $j(x) = x^9 + 4x^5 - x$?

An *even* function is anything you can write using only even powers; an *odd* function is anything you can write with only odd powers. (Some functions – in some sense, most functions – are neither even nor odd.)

However, some functions are even or odd without it being obvious you can write them in the appropriate form: $\sin(x)$ and $\tan(x)$ happen to be odd functions, while $\cos(x)$ is even. Looking at the graphs of a few of these functions gives you an idea of what it might mean to be odd or even – see Figure 9-6.

(a)

(b)

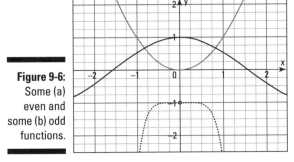
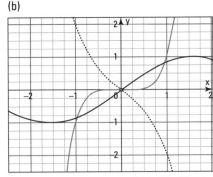

Figure 9-6:
Some (a)
even and
some (b) odd
functions.

© John Wiley & Sons, Inc.

An even function has reflective symmetry in the y-axis; an odd function has rotational symmetry about the origin. There's a more formal definition, though:

- ✔ If $f(-x) = f(x)$ for all x, the function has reflective symmetry in the y-axis, so f is an even function.

- ✔ If $f(-x) = -f(x)$ for all x, the function has rotational symmetry about the origin, so f is an odd function.

- ✔ If neither of these conditions holds, then the function is neither even nor odd.

Numerical Methods

I know, I've been banging on for the whole of the book about how exact answers are brilliant, and you should always use fractions, and decimals are evil. There's *one* topic where that's not true: numerical methods.

The thing is, not every mathematical problem has an answer you can write down exactly (except by cheating – for example, by defining a constant to be the unknown number you want). In these cases, it makes sense to come up with decimal approximations rather than just shrugging and saying, 'There's no answer.'

This is straying dangerously into the world of computer maths, which, obviously, Ofqual insists you attempt without a computer, in the same way that driving tests should be done without a car. No, hang on – that would be ridiculous. Forget I said anything.

In this section, I show you the three kinds of numerical maths you'll be expected to do at A level:

- ✔ **Iteration:** This is basically 'doing the same thing over and over again until it settles down on an answer' – something computers were designed to do.

- ✔ **Root-bounding:** This is trial and improvement with a posher hat on. Again, it's the kind of thing computers were designed for.

- ✔ **Numerical integration:** You can approximate the area under a curve by splitting it into smaller shapes and adding up their areas. You'll never guess what, but computers... no, keep guessing. *Computers* were designed for this.

But for all that computers are kings (or queens) of numerical methods, you need to be able to do iteration, root-bounding and numerical integration on paper. Now go and draw a three-point turn.

Iteration

The word *iterate* means, roughly, 'do again' – and *iteration*, in terms of Core 3, at least, means 'repeatedly feeding a function's result back to the same function in the hope that it will converge'. *Converge*, for now, means 'stop changing'.

For example, suppose $f(x) = \sqrt{\dfrac{2}{x}}$. Work out $f(1)$, which is $\sqrt{2}$. Now put $\sqrt{\dfrac{2}{Ans}}$ into your calculator and press the equals button until your fingers get sore. You'll notice that the answer gets closer and closer to something about 1.26 – it converges on this value.

What you're doing here is solving the equation $x = \sqrt{\dfrac{2}{x}}$, which you can rearrange into $x^2 = \dfrac{2}{x}$, or $x^3 = 2$. (Your calculator will happily tell you that $\sqrt[3]{2} \approx 1.2599$.)

Iteration questions usually consist of three parts:

✔ Rearranging a formula into a specified form

✔ Iterating from an initial guess until the answer converges

✔ Showing that the answer is correct to a specified accuracy

In this section, I take you through the first two parts, and I cover the last part in the next section, 'Root-bounding'.

Rearranging a formula for iteration

For example, a question might define a function as $f(x) = -x^3 + 5x - 3$ for $x > 1$ and ask you to show that $f(x) = 0$ can be rearranged as $x = \sqrt{5 - \dfrac{3}{x}}$.

Notice that there's an x on both sides of the equation here; this is an exercise in rearrangement rather than solving for x. Here's how to do it:

1. **Write $f(x) = 0$ explicitly.**

   $$-x^3 + 5x - 3 = 0$$

2. **Decide which x you're going to try to isolate.**

   Because there's no x^3 in the answer, it makes sense to work with the x^3.

3. **Rearrange until you get that x on its own.**

   $x^3 = 5x - 3$, so $x^2 = \dfrac{5x - 3}{x}$ and $x = \sqrt{5 - \dfrac{3}{x}}$. (You can ignore the negative square root since you know x is greater than 1.)

Don't worry too much about getting the right x in Step 2 – if you've been working at it for a while and don't seem to be getting close, try to isolate a different x! Also, you can use the answer you're aiming for as a guide – your answer doesn't have a cube root in it, but it does have a square root, so you presumably need an x^2 on its own at some point.

An alternative method that almost always works is to start from the target and try to work backwards to the original function. Do this on a spare page somewhere out of sight, and then go back through your steps in reverse order. It's sneaky – but it works!

Iterating with a calculator

In the second part of this question, you'd be asked to use an initial estimate of $x_0 = 1$ and the iterative scheme $x_{n+1} = \sqrt{5 - \dfrac{3}{x_n}}$ to find x_5, an approximate solution to $f(x) = 0$. What you do is this:

1. **Work out the right-hand side using x_0.**

   I get $\sqrt{2} \approx 1.414$, which is x_1. Write this down.

2. **Type the right-hand side into your calculator, but replace x_n with *Ans*.**

   Pressing '=' will give you x_2, which is approximately 1.697. Using the 'Ans' button at each iteration means the calculator uses the exact (or at least, very precise) answer it's worked out rather than this rounded version.

3. **Keep pressing '=' and writing down the result until you reach the subscript they want.**

   Here, $x_3 = 1.798$, $x_4 = 1.825$ and $x_5 = 1.832$, all given to three decimal places although calculated to many more. Obviously, if the exam says 'give each result to five decimal places', then you should give each result to five decimal places; if it's not stated, four significant figures is plenty.

This answer turns out to be correct to two decimal places – the actual solution is around 1.834.

Using your calculator efficiently saves you a great deal of time in the exam – the 'Ans' button (which the calculator mentally replaces with the last thing it worked out) is your friend.

Root-bounding

If, one minute, you see a chicken on one side of the street and then, a few minutes later, the chicken is on the other side, what do you conclude? Obviously, it's crossed the road in the meantime. (Personally, I question its motives.)

You can use similar reasoning to show that an equation has a solution between two given values or that the solution is correct to a certain number of decimal places. To show that the solution from the preceding section ($x_5 = 1.832$) is correct to two decimal places, here's what you do:

1. **If you're not given explicit limits, work them out.**

   Here, you're working to two decimal places; for 1.83 to be correct to two decimal places, the real answer must be $1.825 \le x < 1.835$.

2. **Rearrange the equation so you have an expression equal to 0.**

   This step isn't strictly necessary, but it's a good habit; with this one, you can go back to the original problem, which is to solve $-x^3 + 5x - 3 = 0$. Your function is the left-hand side, $f(x) = -x^3 + 5x - 3$.

3. **Work out the value of the expression for your lower limit, and make a note of it.**

   Here, $f(1.825) \approx 0.047$.

4. **Work out the value of the expression for your second value, and make a note of it.**

   You get $f(1.835) \approx -0.0039$.

5. **If the values have different signs, congratulations! Your solution is clearly between the limits you had.**

You can use an identical method to show that a root is correct to (say) three decimal places. Here's how:

1. **Find the upper bound of your rounded answer.**

   If you want to show the solution 3.655 is correct to three decimal places, the actual solution has to be below 3.6555 (the smallest number that would round up to 3.656).

2. **Find the lower bound of your rounded answer.**

   In this example, you know the solution needs to be at least 3.6545 (the smallest number that rounds up to 3.655).

3. **Apply the previous recipe to show the root lies between these values.**

   Whatever value the root has, if it lies between these two values, it must be 3.655 to three decimal places.

Don't give them any of that 3.65549̇ nonsense for the upper bound. Yes, it's strictly the same thing as 3.6555, but in that case, why not just write 3.6555? 3.6554 is right out.

Integrating numerically

In the real world, it's not always possible (or desirable) to integrate everything *analytically*; instead, it's often simpler to throw computing power at a problem and get a good approximation quickly rather than an exact answer in a few months, if ever.

Unfortunately, in an exam, you don't have (much) computing power to throw at questions – although your calculator is probably more advanced than the computers on the Apollo space missions. The upshot is that the exams are a bit limited in how much numerical integration they can ask you to do.

In most of the exam boards, numerical integration consists of the trapezium rule and nothing more. However, some boards (notably OCR) do ask about Simpson's rule, which is more involved but gives better approximations.

Trapezium rule

The idea of the *trapezium rule* is simple: instead of bothering with all that fiddly integration nonsense, how about using the curve itself to define a series of trapeziums – and then use the total area of the trapeziums to estimate the area under the curve?

It's a good way to do things: if you use enough trapeziums, you can get as close as you like to the real area of any sensible curve. In an exam, though, you'd only be expected to work with maybe five or six trapeziums, at the outside; and you have the formula in the book. For reference, though, the trapezium rule is

$$\int_a^b y\, dx \approx \frac{1}{2}h\left(y_0 + y_n + 2\left(y_1 + y_2 + \ldots + y_{n-1}\right)\right)$$

Here, n is the number of trapeziums you're using (1 fewer than the number of points) and h is the horizontal 'height' of each trapezium – which works out to be $\frac{b-a}{n}$. (The height is horizontal because the parallel sides of each trapezium are vertical.) All those ys, known as *ordinates*, are the values of the function you work out at equally spaced x-values. Here's a recipe for working out $\int_0^1 e^{x^2}\, dx$ using four trapeziums (or, equivalently, five ordinates):

1. **If you're not given a table, make one like the one in Figure 9-7.**

   It needs columns for x, y, a multiplier m and a result, my.

2. **In the x column, work out the equally spaced x-values you need.**

   In this case, you'll have five (it's always 1 more than the number of trapeziums), at 0, 0.25, 0.5, 0.75 and 1.0. Make a note of h, the difference between them: it's 0.25.

3. **Work out the y-value at each of these points.**

   The values are 1, 1.064, 1.284, 1.755 and 2.718 (to three decimal places). Fill these in.

4. **In the multiplier column, put '1' in the first and last rows and '2' everywhere else.**

   This reflects the rule you're given: the first y-value (y_0) and the last (y_n) show up once, and everything else is doubled.

5. **Multiply each y-value by its multiplier and put the result in the final column.**

   I get 1, 2.129, 2.568, 3.510 and 2.718. (I've used the exact value here rather than the rounded one; either is fine in the exam as long as you're consistent about it.)

6. **Add these up and multiply by $\frac{1}{2}h$.**

   I get $11.925 \times \frac{1}{2} \times 0.25 \approx 1.491$ to three decimal places, which is my answer.

x	y	m	my
0	1.000	1	1.000
0.25	1.064	2	2.129
0.5	1.284	2	2.568
0.75	1.755	2	3.510
1	2.718	1	2.718
			11.925
		$\times \frac{0.25}{2} = 1.491$	

Figure 9-7: The trapezium rule in action.

© John Wiley & Sons, Inc.

As a sanity check, I'd ask what the area of the 'big trapezium' was, just using y_0 and y_4: it's 1 unit wide, 1 tall on the left and a bit less than 3 units tall on the right, so it would have an area of about 2. Because the graph is below the trapezium throughout, something a bit below 2 seems quite reasonable.

Simpson's rule

Simpson's rule is very similar to the trapezium rule, except instead of using trapeziums, it fits a series of quadratics to the curve. It's a nice exercise to try to work out the precise reasons it works, but I leave that for the interested reader.

Again, should you need it, the formula is in the book:

$$\int_a^b y\,dx \approx \frac{1}{3}h\left(y_0 + y_n + 4\left(y_1 + y_3 + \ldots + y_{n-1}\right) + 2\left(y_2 + y_4 + \ldots + y_{n-2}\right)\right)$$

This formula looks more complicated than the trapezium rule, but the method is just the same – except for the multiplier column. (Instead of $1, 2, 2, \ldots, 2, 1$, it's $1, 4, 2, 4, 2, \ldots, 2, 4, 1$).

For Simpson's rule to work, you need n to be even, making $n+1$ points altogether; the odd-numbered y-values in the middle will be multiplied by 2, and the even-numbered terms, by 4.

Like with the trapezium rule, h is the horizontal distance between your x-values; unlike the trapezium rule, you divide by 3 instead of 2 at the end. Using the same example as before, here's what you do:

1. **If you're not given a table, make one like the one in Figure 9-8.**

   It needs columns for x, y, a multiplier m and a result, my.

2. **In the x column, work out the equally spaced x-values you need.**

   In this case, you'll have five (it's always 1 more than the number of strips), at 0, 0.25, 0.5, 0.75 and 1.0. Make a note of h, the difference between them: it's 0.25.

3. **Work out the y-value at each of these points.**

   They're 1, 1.064, 1.284, 1.755 and 2.718. Fill these in.

x	y	m	my
0	1.000	1	1.000
0.25	1.064	4	4.258
0.5	1.284	2	2.568
0.75	1.755	4	7.020
1	2.718	1	2.718
			17.564
		$\times \dfrac{0.25}{3} = 1.464$	

Figure 9-8:
Simpson's
rule.

© John Wiley & Sons, Inc.

4. In the multiplier column, put '1' in the first and last rows, and alternate between '4' and '2' everywhere else.

This reflects the rule you're given: the first y-value (y_0) and last (y_n) show up once, odd ys are multiplied by 4, and even ys are multiplied by 2.

5. Multiply each y-value by its multiplier and put the result in the final column.

I get 1, 4.258, 2.568, 7.020 and 2.718, again reported to three decimal places but using the full answer.

6. Add these up and multiply by $\frac{1}{3}h$.

I get $17.564 \times \left(\frac{1}{3} \right) \times 0.25 \approx 1.464$, which is my answer.

This is in the same ballpark as the trapezium rule answer, but it's a bit more accurate; the true answer is around 1.463, which you could reach by using a bigger n.

If you're told the number of ordinates to work with, your number of strips or trapeziums, n, is always 1 fewer. A question asking you to work with seven ordinates is the same as one asking you to work with six strips.

Part III
Geometry

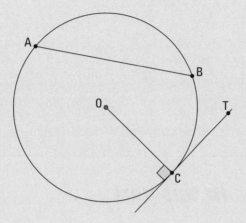

There's much more to geometry than this part of the book has a chance to dig into —
but luckily, there's more space online! Visit www.dummies.com/extras/
asalevelmathsuk for an in-depth explanation of why Bad Guy *x* behaves the way
he does and why the grown-up equation of a line is better.

In this part . . .

- ✔ Get to grips with coordinate geometry.
- ✔ Square up to circles and triangles.
- ✔ Take on the tricks of trigonometry.
- ✔ Vanquish vectors.

Chapter 10

Coordinating Your Geometry

. . . .

In This Chapter

▶ Writing the equations of a line

▶ Sketching and transforming graphs

▶ Finding where curves meet axes – and each other!

. . . .

I had lunch with a university lecturer a few weeks ago, and I asked him about the biggest gap students had in their knowledge. Good news: he said that students with A levels tend to be on top of things! There was a *but*, though: he said that the one problem was the link between algebra and geometry. Students – in his experience – struggle to use algebra to think about curves and vice versa.

This chapter is the one where I try to put that right with you. Equations of a line? You'll be making mincemeat of them by the end of the chapter. (Mmm, mincemeat.) Sketching graphs? Bad Guy *x* and Good Guy *y* will make them a piece of cake. (Mmm, cake.) And as for working out where curves meet the axes and other curves – that's going to be as easy as pie. (Mmm, pie.)

Excuse me. I'm going to go and grab a bite to eat.

The Many Equations of a Line

You probably – hopefully – have an idea of what a straight line is. Mathematically, it's defined as 'what joins two points' – and it goes on forever in both directions.

As far as A level goes, you only have to deal with two-dimensional lines (apart from vector lines, which crop up in Core 4 – and Chapter 13). Two-dimensional lines have a *gradient* – a measure of steepness defined as how far you move in the *y*-direction for every unit you move in the *x*-direction. Or, in English, how far up you go for every one you go to the right. That's a critical part of the equation of a line, whichever one you use: in this section, you see three possibilities for writing it and several for working out the gradient.

Finding a gradient

Depending on the question, there are several ways of getting the gradient of a line, any one of which you might need to know. Here are the scenarios I can think of:

- ✔ You're given the gradient, or, less straightforwardly, you're given the equation of a line that's parallel or perpendicular.
- ✔ You're given two points on the line.
- ✔ You're told the line is a tangent to a circle.
- ✔ You're told the line is a tangent or normal to a curve.

In this section, I give you a recipe for each scenario.

Given the equation of a parallel or perpendicular line

If you're given the equation of a line that's parallel or perpendicular to the line you want, follow these steps:

1. **Rearrange the equation to get *y* on its own. The gradient of the given line is the number in front of the *x*.**

   For example, if your given line is $2x + 3y + 5 = 0$, you get $y = -\frac{2}{3}x - \frac{5}{3}$. It's much better to separate the terms like this rather than write them over a single denominator; something like $y = \frac{-2x-5}{3}$ is trickier to interpret. Written the first way, it's clear this line has a gradient of $-\frac{2}{3}$.

2. **Determine the gradient of the other line:**

   - If your line is parallel to the first one, that's your gradient, too.
   - If your line is perpendicular to the first, your gradient is the negative reciprocal of that.

   The parallel line also has a gradient of $-\frac{2}{3}$.

   To get the negative reciprocal, change the sign and then, if the gradient is a fraction, turn it upside down; if it's a whole number, divide 1 by that number. Here, you get $\frac{3}{2}$ as the gradient of the perpendicular line.

Given two points on the line

If you know two points lie on a line, you can draw a right-angled triangle with a horizontal base and a vertical side so that the part of the line that's between the two points is a hypotenuse. You don't care how long the hypotenuse is (although if you did, you could ask Mr Pythagoras for some help); what you care about is the *rise* (the length of the vertical side, with a minus sign if the line's going downhill) over the *run* (the length of the horizontal side), which you can see in Figure 10-1. In that example, the rise is 5 and the run is 3, so the gradient is $\frac{5}{3}$.

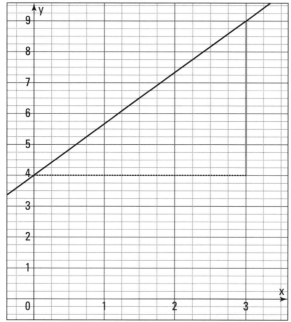

Figure 10-1:
A straight
line.

If the line is going up and to the right, you should get a positive gradient; if it's going down and to the right, it should be a negative gradient.

Tangent to a circle

If you're asked for the gradient of a line tangent to a circle, you'll be given a point on the edge of the circle, and you'll have already worked out the centre of the circle. (If you haven't, you should; if you don't know how, check out Chapter 11.)

The line between those two points (the centre and the given point) is a radius, and the tangent is perpendicular to that. You can find the gradient of the radius using the recipe for two given points (see the preceding section); then take the negative reciprocal to find the tangent's gradient.

Tangent or normal to a curve

If your line is tangent to a curve at a given point, the line has the same gradient as the curve does there – which you can get by differentiating the curve's equation and substituting in the known *x*-value.

For a normal, you do the same thing, only you take the negative reciprocal of what you get.

A *tangent* to a curve at a given point is the straight line through the point that has the same gradient as the curve; a *normal* to a curve at a given point is the straight line through the point that's perpendicular to the tangent.

I go into tangents and normals in *much* more detail in Chapter 15.

The one you know: $y = mx + c$

After you know a gradient, you can use the old familiar $y = mx + c$ to find the equation of the whole line. Here's how you find the equation of a line with gradient $\frac{3}{5}$ through the point $(7, 1)$:

1. **Replace m with your gradient, x with your known x-value and y with your known y-value.**

   Here, you get $1 = \frac{3}{5} \times 7 + c$.

2. **Rearrange to get c.**

$$c = 1 - \frac{21}{5}$$
$$= -\frac{16}{5}$$

3. **Write down your final equation.**

$$y = \frac{3}{5}x - \frac{16}{5}$$

Delightful as all those fractions are, it's unusual (but far from unknown) for an A level question to ask for the equation in this form. The examiners usually much prefer $ax + by + c = 0$ – which is easier to reach from the equation in the next section.

The better one: $\left(y - y_0 \right) = m \left(x - x_0 \right)$

I don't usually say, 'This equation is better than that one' – different strokes for different folks and all that – but the equation $\left(y - y_0 \right) = m \left(x - x_0 \right)$ is better than $y = mx + c$. No contest, especially when you have a fractional gradient, a question that wants $ax + by + c = 0$ as its final form, or (as is really common in A level) both. Here's how it works:

1. **Replace *m* with the gradient, and replace x_0 and y_0 with your known coordinates.**

   For a line with gradient $\frac{3}{5}$ through the point $(7, 1)$ – the same example as before – you have

   $$(y - 1) = \frac{3}{5}(x - 7)$$

2. **Multiply both sides by the bottom of the fraction (if needed) and expand.**

   $$5y - 5 = 3x - 21$$

3. **Rearrange into the form they want.**

   $$3x - 5y - 16 = 0$$

That's it! No messing about with fractions (usually), beyond multiplying up to get rid of them – and the correct form just drops out of the working.

This form is also closely linked to the definition of the gradient. You can easily rearrange $(y - y_0) = m(x - x_0)$ to $m = \frac{y - y_0}{x - x_0}$ – or, if you prefer, the rise over the run. It's simple, obvious and easy to use, and I say again: it's better than $y = mx + c$. So there.

Sketching with Skill

It's a rare exam paper that doesn't ask for some kind of graph sketch. Even if you're not asked explicitly for a sketch, drawing out what's going on can often give you insight into what you're looking for.

So grumble all you like: sketching is an important part of your toolkit as a mathematician. Here are my Sketching Rules:

- ✔ **Draw big.** Make sure you leave plenty of room for labels.

- ✔ **Draw in pencil.** You'll make mistakes, and it's easier to rub things out and put them right than it is to redo the whole picture.

- ✔ **Don't worry too much about accuracy.** You want your picture to be the right sort of shape and for everything to labelled clearly; if you wanted an accurate plot, you'd fire up a spreadsheet or desmos.com.

Starter kit: Getting the basic shapes right

There are several graph shapes you need to be able to sketch, which I've split up into 'basic' – the ones you'll come across in Core 1 and 2 – and 'advanced' – which you'll need for Core 3 and 4. You'll be expected to be able to transform these (see 'Tricky Transformations' later in this chapter) and make deductions from them.

It's OK if the shape of your graph isn't bang-on perfect – your quadratics don't have to be pinpoint accurate, and your axes don't need to be meticulously labelled with numbers – or even have the same scales. As long as your graphs bear a passing resemblance to what they're meant to look like, you'll be fine.

The basic graphs

The first graphs you need to know about are the polynomial graphs – the straight line (Figure 10-2a), the quadratic (Figure 10-2b) and the cubic (Figures 10-2c and 10-2d):

- ✔ **Straight line:** For a straight line, you need to make sure the gradient looks reasonable and that the line crosses the axes in the right places.

- ✔ **Quadratic:** A quadratic needs to be the right way up (if the number of x^2s is positive, it's a smiley face; if the x^2 coefficient is negative, it's a frowny face) and cross the axes in the right places.

- ✔ **Cubic:** The cubic also needs to be the right way up (if the number of x^3s is positive, the cubic goes the same way as a positive-gradient straight line – from the bottom left to the top right). Again, the graph needs to cross the axes in the correct place(s).

You'll also need to deal with *reciprocal* graphs, such as $y = \dfrac{1}{x}$ (also written $y = x^{-1}$, if you recall your power laws from Chapters 3 and 5) and $y = \dfrac{1}{x^2}$ (or $y = x^{-2}$). In both cases, you have a problem when $x = 0$: you can't divide by 0!

You can see what happens if you think about numbers just above and just below 0 – for $y = \dfrac{1}{x}$, if x is a tiny positive number, you get a huge positive number out; if x is a negative number very close to 0, your y is a negative number very far away. You can see the graph's behaviour in Figure 10-3a – the graph goes off to infinity, giving you a *vertical asymptote* at $x = 0$. An *asymptote* (in A level, at least) is a line to which a curve can get as close as you like. There's also a *horizontal asymptote* at $y = 0$, as the graph gets as close as you like to the *x*-axis.

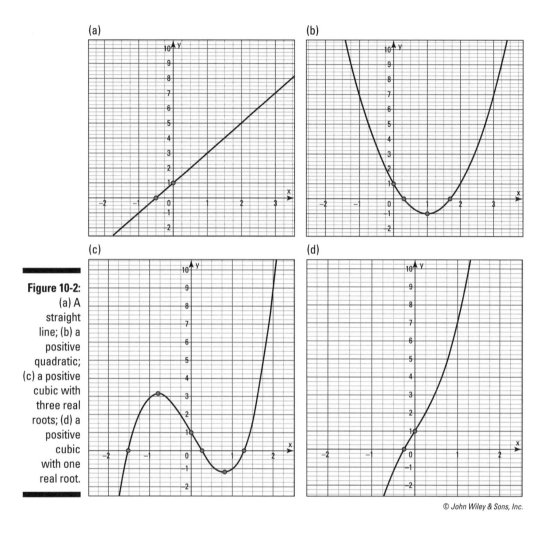

Figure 10-2: (a) A straight line; (b) a positive quadratic; (c) a positive cubic with three real roots; (d) a positive cubic with one real root.

There's a bit more to the definition of an asymptote; the following definition isn't perfectly precise, but it gives you enough of a flavour to go on with. A vertical line $x = a$ is an asymptote if, immediately to the left or right of the line, the graph 'goes to infinity' vertically – that is to say, the y-value gets larger in magnitude than any number you can think of. Similarly, a horizontal line $y = b$ is an asymptote if, immediately above or below the line, the graph 'goes to infinity' horizontally. In both cases, the curve skates along the asymptote.

Contrary to popular belief, a curve *can* intersect an asymptote, but it doesn't have to. For example, the curve $y = \dfrac{\cos(x)}{x}$ has a horizontal asymptote at $y = 0$, although it crosses that line infinitely often.

If $y = \dfrac{1}{x^2}$ and if x is very close to 0, then y is huge and positive, regardless of whether x is positive or negative (it's squared, so it becomes positive). The asymptotes are the same as for $y = \dfrac{1}{x}$ (with a vertical one at $x = 0$ and a horizontal one at $y = 0$), but the shapes are different – this graph is always above the x-axis, as you can see in Figure 10-3b.

(a)

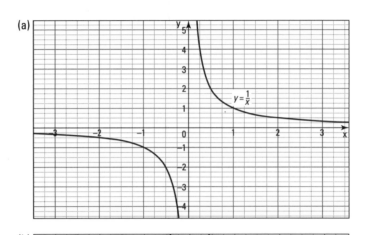

(b)

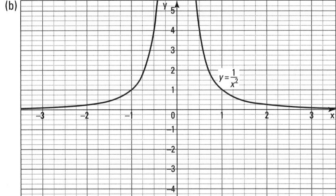

Figure 10-3:
(a) $y = \dfrac{1}{x}$
and
(b) $y = \dfrac{1}{x^2}$.

© John Wiley & Sons, Inc.

Basic trigonometry graphs

It's unusual to have to sketch the sine, cosine or tangent graphs in Core 2, but it's a *really useful skill to have*. Knowing the shapes and symmetries of the three graphs makes it much easier to spot where you have multiple solutions to a trig question and which answers make sense.

Figure 10-4a shows the sine graph, which goes through $(0, 0)$; one of my students told me to remember it as 'sine starts in the sea', which is more memorable than anything else I've ever heard. From there, interesting things happen every $\frac{1}{2}\pi$ radians (or 90°). Moving to the right, the curve peaks at $\left(\frac{1}{2}\pi, 1\right)$; then it crosses the axis again at $(\pi, 0)$, hits a minimum at $\left(\frac{3}{2}\pi, -1\right)$, and returns to the axis at $(2\pi, 0)$. You could, of course, do that in degrees, but why would you, unless the question asked for it?

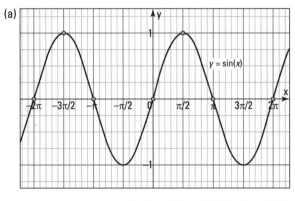

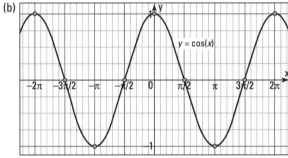

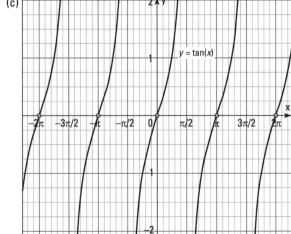

Figure 10-4: The basic (a) sine, (b) cosine and (c) tangent graphs.

The cosine graph is in Figure 10-4b. It goes through $(0, 1)$; the mnemonic says 'cos starts in the clouds'. Again, interesting things happen every $\frac{1}{2}\pi$ radians: at $\left(\frac{1}{2}\pi, 0\right)$, the curve crosses the axis; the cosine curve then hits its minimum at $(\pi, -1)$ before crossing the axis again at $\left(\frac{3}{2}\pi, 0\right)$ and reaching a maximum at $(2\pi, 1)$.

The graph of $y = \tan(x)$ is the odd one out; it doesn't have the nice wavy behaviour of sine and cosine (although it is *periodic,* meaning the same pattern repeats forever). The graph goes through $(0, 0)$, rising up to a vertical asymptote at $x = \frac{1}{2}\pi$. It comes back in from below to cross the axis at $(\pi, 0)$ and rises to another vertical asymptote at $x = \frac{3}{2}\pi$ before coming back to the axis at $(2\pi, 0)$.

Why does the tangent behave in this pathological way? Simple: $\tan(x)$ is identical to $\dfrac{\sin(x)}{\cos(x)}$, so it's 0 whenever $\sin(x) = 0$ and undefined whenever $\cos(x) = 0$.

Trickier shapes: Sketching the advanced graphs

Just when you think you're on top of sketching graphs in your AS level, Core 3 suddenly asks you to deal with all manner of new functions: exponentials and logarithms (Chapter 5 has more on these) and the minor trig functions (secant [or sec], cosecant [or cosec], and cotangent [or cot], which you find in Chapter 12). There's also the modulus function, which I'm arbitrarily going to treat as a transformation rather than a function in its own right – see the section 'The madness of the modulus' later in this chapter for details.

You can try to remember all the shapes – and it's not a bad idea – but as long as you can *generate* the graphs as needed, you're on solid ground.

Exponential and logarithmic graphs

Exponential and logarithmic graphs (as shown in Figure 10-5) are reflections of each other in the line $y = x$, because they're inverses of each other. (If you've read up on functions and inverses in Chapter 9, you know this already.) I've picked e as my base, because it's the best base, as Chapter 5 tells you.

The exponential graph $y = e^x$ crosses the y-axis at $y = 1$ because $e^0 = 1$. To the right, it gets very big very quickly: e^3 is already over 20, and e^6 is somewhere around 400. The flip side of that is, to the left, it gets very small very quickly but stays (just) positive: e^{-10} is a very small positive number. To the left, the graph has a horizontal asymptote at $y = 0$.

By contrast, the logarithmic graph $y = \ln(x)$ is defined only for positive values of x. It crosses the x-axis at $x = 1$ (because $\ln(1) = 0$) and then grows quite slowly: $\ln(60)$ is only a little bit over 4. Between 1 and 0, it reaches a very big negative number very quickly, and the graph has a vertical asymptote at $x = 0$.

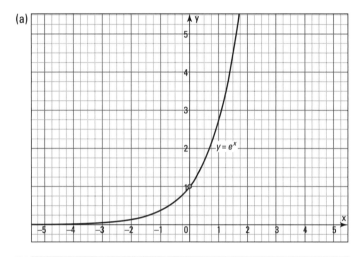

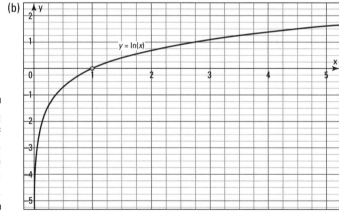

Figure 10-5: Graphs of (a) exponential and (b) logarithmic functions.

© John Wiley & Sons, Inc.

These shapes work for any number bigger than 1 in place of the e – the graphs of $y = 1.1^x$ and $y = 17^x$ look just the same as the e^x graph, only stretched or shrunk as appropriate (see the section 'Looking out for Bad Guy X' later in this chapter). If the base of an exponential function is between 0 and 1, the graph will be reflected in the y-axis – which makes sense if you think about it: as the power increases, the y-value gets smaller and smaller.

Similarly, the graph of $y = \log_a(x)$ is a vertical stretch of the natural log graph if $a > 1$, and it's a stretch combined with a vertical reflection in the x-axis if $0 < a < 1$.

Minor trig functions

The minor trig functions – $\sec(x)$, $\csc(x)$ and $\cot(x)$ – are the reciprocals of $\cos(x)$, $\sin(x)$ and $\tan(x)$, respectively. You use that fact when constructing their graphs, which are in Figure 10-6.

Here's a recipe for drawing a minor trig function graph (this is a pretty good method for drawing *any* reciprocal graph, by the way!):

1. **Lightly sketch the major trig function that goes with it.**

   For $y = \sec(x)$, you'd sketch $y = \cos(x)$.

2. **Anywhere your major curve reaches $y = \pm1$, mark a point.**

   A maximum on a major curve will become a minimum on the minor curve and vice versa. If you don't have a turning point at $y = \pm1$ on your original curve, you won't have one on the reciprocal curve, either.

3. **Anywhere your major curve reaches $y = 0$, draw a vertical asymptote.**

   You weren't about to try to divide by 0, were you? Naughty, naughty.

4. **Anywhere your major curve has a vertical asymptote, draw a point at $y = 0$.**

   This step applies only if you're drawing $y = \cot(x)$.

5. **Join the dots!**

   $y = \sec(x)$ and $y = \csc(x)$ have a distinctive bucket-and-stool shape, while $y = \cot(x)$ is a reflection of $y = \tan(x)$ in the line $x = \frac{\pi}{4}$ (among others).

Intercepts: Crossing your xs and ys

If you have the equation of a graph, finding where it crosses the y-axis is as simple as substituting $x = 0$ in and finding what comes out as your value for y. The huge majority of graphs you see at A level won't even need rearranging.

(a)

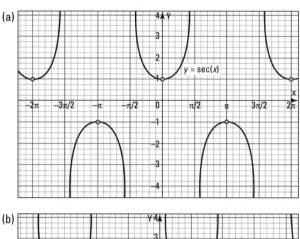

(b)

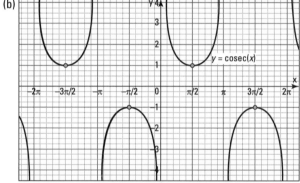

(c)

Figure 10-6:
Graphs of
the minor
trig func-
tions: (a)
secant, (b)
cosecant
and (c)
cotangent.

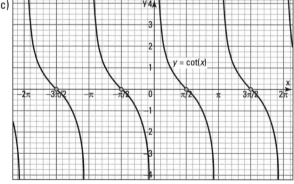

© John Wiley & Sons, Inc.

If you're dealing with a function (and you almost always are – circles, para-
metrics and implicits aside), you'll have only one solution. In those cases,
answering the problem is just a case of rearranging what you have left to
solve for *y*.

The method for finding where the graph crosses the x-axis, on the other hand, is a little trickier: you need to substitute in $y = 0$ and then solve for x. You should be on the lookout for several answers in most cases.

Covering your asymptotes

Finding a vertical asymptote is as simple as finding out where your function becomes undefined – either because you're trying to take the logarithm of 0 (rare) or because you're trying to divide by 0 (very common).

You don't get a vertical asymptote when the argument of a square root becomes negative, because the graph of the function simply stops at 0 rather than going to (positive or negative) infinity.

Sometimes the division by 0 is hidden away in a tangent or in a minor trig function. Because $\sec(x) \equiv \dfrac{1}{\cos(x)}$, for example, the secant graph has asymptotes wherever $\cos(x)$ has zeroes. Be on your toes if you're sketching such a graph!

Horizontal asymptotes are easier: just imagine (or, if you're admirably lazy, calculate) what happens if x is an enormous positive or negative number.

Sketching with the DATAS method

Putting all this together to give the best possible sketch of a function (the kind of thing you might be asked about in an interview), you can use the mnemonic DATAS to remember what you need to put into a graph. Here's what that stands for:

- **Domain:** Where is the thing you're plotting defined? For example, if you've got a square root hanging about, you know that can't have a negative argument. Similarly, if there's a $\log_a(x)$ or a $\ln(x)$, you know it'll work only if x is positive.

- **Axes:** Where (if anywhere) does your curve cross the x- and y-axes? You can find the y-intercept of the curve (where it crosses the y-axis) by substituting $x = 0$ into your equation, and you can find the x-intercept or intercepts by solving for where $y = 0$.

- **Turning points:** Where (if anywhere) does your curve have a stationary point? What type(s)? You can get these by differentiating and solving for where $\dfrac{dy}{dx} = 0$. See Chapter 15 for more on turning points!

- **Asymptotes:** Where (if anywhere) does your function go off to positive or negative infinity? You can find these places by seeing if the bottom of a fraction ever becomes 0 or if the argument of a logarithm is ever 0.

Look at the behaviour on each side of the asymptote, too – is the function defined? Is it positive or negative?

You also want to think about horizontal asymptotes – what happens if x gets enormous? What if it's the negative of an enormous number?

✔ **Shape:** On the face of it, shape is the trickiest thing – there are countless types of graph, and trying to remember them all is a losing battle. However, if you've got the four preceding points straight, getting the right shape is usually just a case of joining the dots.

Tricky Transformations

In this section, I introduce you to two very helpful chaps, Bad Guy x and Good Guy y. The pair of them, I have to say, aren't the most mathematically rigorous fellows in the world, but they get the job done as far as you're concerned. (If you're interested in why this method works, you can check out the extra material on it at www.dummies.com/extras/asalevelmaths.)

The general principle is that y does what it is told; x does the _opposite_ of that.

I also run you through the modulus function transformation, which doesn't have a character that goes with it, so you'll just have to remember how it works.

Looking out for Bad Guy x

Bad Guy x is a _very_ bad guy. He's in bracket prison because he always does the opposite of what he's told. For example, if you have a graph $y = f(x)$ and you politely ask Bad Guy x to move three spaces to the right by replacing x with $x + 3$, he says, 'Nut-uh! I'm Bad Guy x!' and promptly moves the graph three spaces to the _left_. The same goes for whatever you ask him to do:

✔ **Shifting left or right:**

- You ask him to go three to the right by asking for $f(x + 3)$: the graph moves three to the left – a translation of three units in the negative x-direction, if you're playing 'talk like an examiner' (which you should be).

- You ask him to go four to the left by asking for $f(x - 4)$: the graph moves four to the right – a translation of four units in the positive x-direction.

✔ **Stretching or compressing horizontally:**

- You ask him to quintuple the width of the graph by asking for $f(5x)$: the graph becomes a fifth as wide as it was – an enlargement of scale factor $\frac{1}{5}$ parallel to the x-axis.

- You ask him to halve the width of a graph by asking for $f\left(\frac{1}{2}x\right)$: the graph doubles in width – an enlargement of scale factor 2, parallel to the x-axis.

✔ **Flipping over the y-axis:** The only thing Bad Guy x does right is turn around: $f(-x)$ gives the reflection in the y-axis.

If you ever want to transform a graph horizontally, think about what's happening to x and do exactly the opposite!

You sometimes need to be on your toes for implied brackets – for example, if $y = e^{2x+3}$, the $2x+3$ is grouped together and done before the powering – which means it must be between brackets you can't see. Similarly, with $y = \dfrac{1}{x-2}$, the $x-2$ is in implied brackets, which needs to be worked out before doing the division. In these cases, Bad Guy x has mysteriously been moved to an open prison, but he still does the wrong thing.

Making friends with Good Guy y

Good Guy y, on the other hand . . . well, he's a really good guy! He does exactly what you tell him to – as long as he's *outside* of the bracket-prison. If you have the graph $y = f(x)$, then the following applies:

✔ **Shifting up or down:**

- You ask him to go up two units by asking for $y = f(x)+2$: he goes up two! A translation of two units in the positive y-direction.

- You ask him to go down three units by asking for $y = f(x)-3$: he goes down three – a translation of three units in the negative y-direction.

✔ **Stretching or compressing vertically:**

- You ask him to be twice as tall by asking for $y = 2f(x)$: he stretches as requested – an enlargement of scale factor 2, parallel to the y-axis.

- You ask him to be a third of the height by asking for $y = \frac{1}{3}f(x)$: he shrinks – an enlargement of scale factor $\frac{1}{3}$ parallel to the y-axis.

✔ **Flipping over the x-axis:** You ask him to flip vertically by asking for $y = -f(x)$: the graph reflects in the x-axis.

Transforming a graph vertically is as straightforward as it looks.

The 'Good Guy y' rule works only when you've worked out y in terms of x. Luckily, all the transformations you need to worry about are easy to arrange this way.

The madness of the modulus

You can read in-depth about the modulus function in Chapter 9, but here, I focus just on the graphs it produces.

The graph $y = |x|$ (the vertical lines mean 'the modulus of') looks like the one in Figure 10-7a, and its basic transformations go just as you'd expect: the lines work like a bracket, so adjusting the x inside gives you Bad Guy x doing the wrong thing (for details, see the earlier section 'Looking out for Bad Guy x'). Figure 10-7b shows the graph of $y = |x - 2|$, which is a translation of two units to the right.

Of course, outside of the lines, it's Good Guy y doing what he's told. Figure 10-7c shows the graph of $y = 2|x|$, which is simply a vertical stretch.

Things get messy when you have something more complicated than a linear function inside the lines. Here are the rules:

- ✔ Putting a function inside the modulus lines turns all its y-values positive. That means anything below the x-axis is reflected up above it, as in Figure 10-8a, which shows $y = |x^2 - 4|$.

- ✔ Putting a function's argument inside the modulus lines is complicated. Wherever the argument is positive, the graph doesn't change at all. Where the argument is negative, the graph is obliterated – and replaced with the y-value corresponding to the positive version of the argument.

For example, Figure 10-8b shows $y = \ln|x - 1|$. For $x > 1$, the graph looks just like $y = \ln(x - 1)$, without the modulus. But for $x < 1$, the thing inside the modulus, $(x - 1)$, is negative and is replaced by the corresponding positive number. In this case, it means the graph is reflected in the line $x = 1$, and you get the funnel shape shown in the figure.

Combining transformations

Combining transformations – at least without a modulus sign in them – is nice and straightforward, because the horizontal and vertical parts don't really affect each other. However, getting things in the right order is still important.

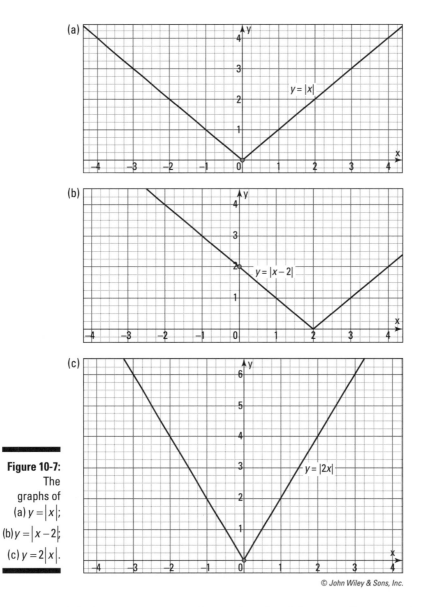

Figure 10-7:
The
graphs of
(a) $y = |x|$;
(b) $y = |x - 2|$;
(c) $y = 2|x|$.

The key thing to remember is this: Good Guy y stuff outside the brackets happens in the order you'd expect from your algebra rules; Bad Guy x stuff happens in the opposite order. Read on to see what I mean!

Simple combinations

A combination where you're simply adjusting both the x and the y, as in $y = 2f(x + 3)$, works nicely: you can do the y-enlargement (scale factor 2) followed by the x-translation (three to the left), or you can do it in the other order – it makes no difference. The problem comes when you need to do two things in the same direction.

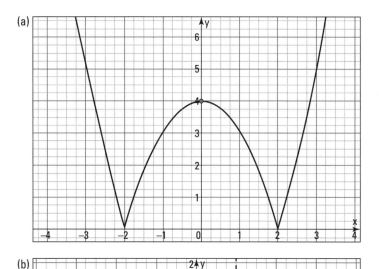

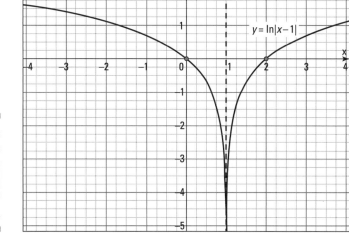

Figure 10-8: The graphs of (a) $y = \left| x^2 - 4 \right|$ and (b) $y = \ln \left| x - 1 \right|$.

However, even a combination like $y = 2f(x) + 3$ obeys the order of operations you know and love: you multiply before you add, so – at least when you're dealing with Good Guy y – you stretch before you move. (***Tip:*** 'Stretch before you move' is also good running advice.)

On the other hand, Bad Guy x always has to be different. If you have $y = f(2x + 3)$, the rule is that you need to do the translation *before* the enlargement. Bad Guy x does everything the wrong way around – even the order of operations!

Another combo: Completing the square

You can see completing the square as a combination of a vertical stretch, a vertical translation and a horizontal translation. In completed square form, you'll have something like $f(x) = a(x-b)^2 + c$ – which means you take the vertex from $(0,0)$ and move it b to the right, stretch the whole curve vertically by the scale factor of a, and then move the curve up c. The vertex ends up at (b, c) – which is exactly what you use completing the square for! For details on completing the square, see Chapter 6.

Combinations with modulus signs

If you've got modulus signs in your transformation, exactly the same rules apply. The graph $y = 2|3x - 6| - 1$ is the classical V-shape of the modulus graph moved six units to the right, shrunk to a third of its width (all the x-coordinates have been divided by 3), doubled in height and moved down one unit – so the point of the V is at $(2, -1)$, the graph crosses the y-axis at $(0, 11)$, and it crosses the x-axis at $\left(\frac{11}{6}, 0\right)$ and $\left(\frac{13}{6}, 0\right)$ – which you might want to verify!

Figure 10-9 illustrates these steps.

Alternatively, you can use the Table mode on your calculator to work out the values of the function at points, just to make sure all is right. (That's a pretty powerful tool for sketching, full stop, but it's nice to try to do it without.) Here's how the Table mode works on my trusty FX-83:

1. **Press MODE at the top of the calculator, followed by whichever number is written next to TABLE.**

   For me, it's 3.

2. **Type in the definition of $f(x)$.**

   There's a modulus bracket in the top left, in case you need it; you can get an x by pressing the red ALPHA button followed by whichever key has a red 'X' above it. For me, that's ')'.

3. **When you're done, press '='. In response to 'Start?', type in the lower end of your domain, followed by '='.**

   If you're not given a domain, –5 is usually fine.

4. **In response to 'End?', type in the upper end of your domain, followed by '='.**

   Generally, 5 is a good default upper limit.

5. **In response to 'Step?', pick a number that's about a twentieth of the difference between your start and end, and press '='.**

If you've used the default values, that's 0.5.

6. **The resulting table lists *x*- and *y*-values for your function between the limits you gave the calculator.**

Plot these, roughly, on your graph to get a sense of the shape.

(a)

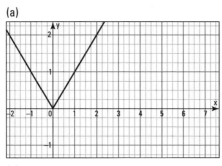

(b)

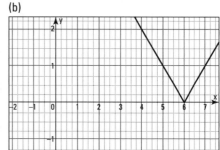

(c)

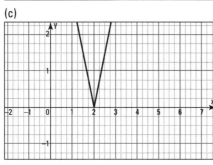

(d)

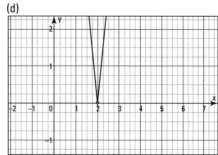

(e)

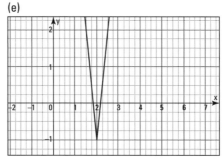

Figure 10-9:
The steps in transforming

$$y = |x|$$

into $y =$

$$2|3x - 6| - 1.$$

Investigating Intersections

A level examiners, in all four of the Core modules, have this insatiable need for very specific knowledge. They're almost obsessed with finding out where curves meet axes, lines and other curves. They want to know if curves cross or merely touch. They want to know, given that they touch in one place, where else they cross.

In short, examiners want you to examine the lives of the curves in extreme, Stasi-like detail and report back with what you've found. In this section, I train you in the methods of the curve-intersection secret police.

Touching and crossing

When curves meet, they can either cross – that is, pass right through each other – or just graze each other. Have a look at Figure 10-10 to see what I mean.

(a)

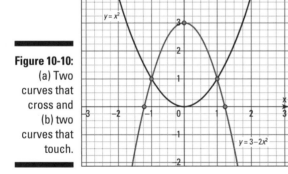

(b)

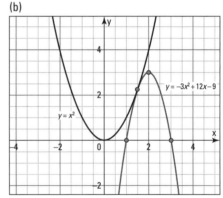

Figure 10-10: (a) Two curves that cross and (b) two curves that touch.

© John Wiley & Sons, Inc.

The pairs of curves I've picked here are $y = x^2$ and $y = 3 - 2x^2$ (which cross twice) and $y = x^2$ and $y = -3x^2 + 12x - 9$ (which just touch).

If you combine the equations in the first case, you get $3x^2 - 3 = 0$, which factorises as $3(x-1)(x+1) = 0$. That has two separate brackets, meaning there are two distinct solutions, at $x = 1$ and $x = -1$. The corresponding y-values are both 1. Neither bracket is raised to an even power here, so your curves cross in both places.

Combining $y = x^2$ with $y = -3x^2 + 12x - 9$, on the other hand, gives you $4x^2 - 12x + 9 = 0$, which factorises as $(2x - 3)^2 = 0$. In that case, you have a single root raised to an even power – so when $x = \frac{3}{2}$, the curves touch rather than cross. The corresponding y-value is $\frac{9}{4}$.

After factorising, the number of distinct factors you have is the number of places where the curves meet; an odd-powered factor gives you a crossing point, and an even-powered factor, a tangent point.

Where curves meet axes

Curves meet the x-axis when $y = 0$ because all the points on the x-axis have a y-coordinate of 0. (If there are multiple solutions at the same point, like you have with $y = (x - 2)^4 (x + 3)$, you need to check whether the curve crosses or touches the axis; the rule is that if the power is even, the curve just touches, whereas for an odd power, the curve crosses the axis.)

Curves meet the y-axis when $x = 0$ because all the points on the y-axis have an x-coordinate of 0.

Sure – that's easy enough to *say*, but what does it mean in practice, especially when you have something more involved than a simple quadratic? In this section, I cover finding the x- and y-intercepts of parametric curves (where x and y are defined in terms of another variable, such as t) and implicit curves (where the xs and ys are all mixed up). You can read more about curves defined parametrically and implicitly in Chapter 18.

Where parametric curves meet the axes

There's an extra step involved in finding where curves defined parametrically meet the axes: as with most things parametric, you need to go through the parameter. If your curve is given by $x = t - 2\sin(t)$ and $y = 1 - 2\cos(t)$ for $0 \le t \le 2\pi$, you can find where the curve crosses either axis by following these steps (I do it for x, but the same method works for y):

1. **The curve crosses the x-axis where $y = 0$, so solve the relevant equation for t.**

   $0 = 1 - 2\cos(t)$, so $\cos(t) = \frac{1}{2}$ and $t = \frac{1}{3}\pi$ or $t = \frac{5}{3}\pi$. If the question just asks for the t-values, you're done here! However, if the examiners want the x-values, carry on.

2. **Substitute this value into the other equation to get the relevant x-values.**

   For the first x-value, $x = \frac{1}{3}\pi - 2\sin\left(\frac{1}{3}\pi\right) = \frac{1}{3}\pi - \sqrt{3}$; for the second,

   $x = \frac{5}{3}\pi - 2\sin\left(\frac{5}{3}\pi\right) = \frac{5}{3}\pi + \sqrt{3}$.

3. If necessary, decide whether the curve crosses or touches the axis at these points.

Neither solution for *t* is an even-powered root, so both answers here are crossing points.

Where implicit curves meet the axes

Solving for where implicit curves meet axes is a lot less complicated than the parametric version. In fact, it's as simple as two bullet points:

✔ To find where an implicitly defined curve crosses the *x*-axis, substitute 0 for *y* and solve the resulting equation for *x*.

✔ To find where an implicitly defined curve crosses the *y*-axis, substitute 0 for *x* and solve the resulting equation for *y*.

If, for example, you have the curve $-2x + y^2 - x^2y = 4 - 2x^2$, you find the *x*-intercepts by substituting 0 for *y* to get $-2x = 4 - 2x^2$, which rearranges to $x^2 - x - 2 = 0$, with solutions at $x = -1$ and $x = 2$.

You find the *y*-intercepts by replacing every *x* with 0 to get $y^2 = 4$, giving you solutions at $y = \pm 2$.

Where curves meet each other

The methods for finding where curves meet each other are very similar to finding where curves cross axes – in fact, the axes are simply the particular curves $y = 0$ (the *x*-axis) and $x = 0$ (the *y*-axis).

The general method is to set up simultaneous equations and solve them. If both curves are explicitly defined in two variables, that's not too tricky: to find where $y = x^2 + 4x + 1$ intersects $y = -2x^2$, you can immediately say, 'I know $y = y$, so $x^2 + 4x + 1 = -2x^2$, which I can turn into a quadratic I can solve.' (For the record, you get $3x^2 + 4x + 1 = 0$, which factorises as $(3x+1)(x+1) = 0$, so $x = -\frac{1}{3}$ or $x = -1$, and the corresponding *y*-values are $-\frac{2}{9}$ and -2.)

Finding where curves meet each other is a little more involved with implicit and parametric equations (see the preceding section) – but not much!

It's unusual for a Core 4 exam to ask you where an implicitly defined curve intersects a (regular) explicitly defined one, although such a problem can come up in Core 2: where does a line (let's say $y = 2x$) intersect a circle (how about $x^2 + y^2 - 6x - 8y = 0$)?

You still have simultaneous equations, which you can solve by substitution: simply replace all the ys in the implicit part with their equivalent values from the first bit – that is, all the ys in the circle equation become $(2x)$s.

The brackets are important here: you need to square and multiply by the whole of y, not just the last bit.

You get $x^2 + (2x)^2 - 6x - 8(2x) = 0$, or $5x^2 - 22x = 0$. That gives you $x = 0$ or $x = \frac{22}{5}$, with the corresponding y-values of 0 and $\frac{44}{5}$ coming from the simpler equation $y = 2x$.

It's a bit more common to figure out where lines cross parametrically defined curves (especially tangents); coincidentally, the process is a bit more involved.

A typical question might give you the line $2x + 3y + 2 = 0$ and the curve C, defined implicitly as $x = t^3$, $y = t^2 - 1$, $t \in \mathbb{R}$. The examiners will tell you (or have you work out) that the line is tangent to the curve at the point where $t = -1$, and they'll ask where the line meets C again. Figure 10-11 shows what it looks like.

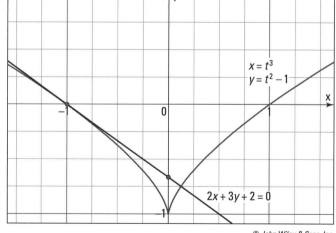

Figure 10-11: A line $2x + 3y + 2 = 0$ intersecting a parametric curve with $x = t^3$ and $y = t^2 - 1$.

The way to work this out is to make use of t by substituting:

1. **Replace the xs and ys in the original equation with their expressions in terms of t.**

   The given equation for the line is $2x + 3y + 2 = 0$, and the curve is defined as $x = t^3$, $y = t^2 - 1$, $t \in \mathbb{R}$. Therefore,

   $$2t^3 + 3\left(t^2 - 1\right) + 2 = 0$$
   $$2t^3 + 3t^2 - 1 = 0$$

2. **Solve this to find all the t-values where the curves coincide.**

   Luckily, you're given the clue that this equation has a solution (in fact, a double root) where $t = -1$, so you can use the factor theorem: $(t + 1)$ is a factor, so divide it out. The equation works out as $(t + 1)^2 (2t - 1) = 0$.

3. **Find the t-value you didn't know about.**

   $t = \frac{1}{2}$ is news to you!

4. **Substitute this value back into the x and y equations to get the point!**

   $x = \frac{1}{8}$ and $y = -\frac{3}{4}$.

Chapter 11

Making Sense of Circles and Triangles

*T*he two most important shapes in A level maths are the triangle and the circle. Of the two, triangles are much more useful, but you still need to know about circles.

In particular, you need to know about measurements of parts of a circle – sectors, chords and arcs – as well as the coordinate geometry of a circle, which means finding its equation or that of tangents, diameters and what-have-you.

Sometimes you'll get shapes that are a mixture of parts of a circle and triangles. Their measurements are also fair game, which means your trigonometry needs to be pretty good.

Oh, and I almost forgot: there's a whole new, infinitely better way of measuring angles than those smelly degrees they made you use at GCSE. The *radian* is the correct way to measure angles. Send anyone who tells you differently my way – they need to be set straight.

Equations of a Circle

At some point in your maths career, you'll try to memorise the equation of a circle. For reference, if your circle's centre is at (a, b) and the circle has radius r, the equation is

$$(x - a)^2 + (y - b)^2 = r^2$$

I don't think you necessarily need to *remember* the equation of a circle – but you do need to be able to work it out if you don't remember it. In this section, I explain it in a way that (I hope) makes remembering it unnecessary! I also show you how to convert between this form and the other, less useful way you might see the equation written; how to deal with simultaneous equations involving circles; and – perhaps most importantly in terms of picking up Core 2 marks – all the geometrical reasoning you need for dealing with tangents, chords and the other odd things they might ask you about.

Where the circle equation comes from

When I teach circles to my students, I usually start by asking, 'What is a circle?' Answers vary. It's a round thing. It's got infinitely many sides. Some will sketch it to demonstrate. It's unusual to get the answer I'm looking for, so there's a follow-up question: how would you draw a perfect circle?

Everyone, correctly, reaches for a pair of compasses.

A *circle,* mathematically speaking, is all the points a certain distance, *r,* from a fixed point, *C.* If you think about it, that's exactly what a pair of compasses gives you: you stick the pointy end in the paper (that's the fixed point, *C,* for centre), stretch the arms apart a distance of *r* (the radius), and spin it around to cover all the points that far away from the centre.

Great. You know what a circle is and how to draw it. Now, how about a mathematical equation for it? No, not πr^2 – that's the area. I want to know whether a general point (x, y) is on a circle – let's say with a centre at (a, b).

What do you know about the point (x, y)? It's a distance r from (a, b), so you can use Pythagoras's theorem to state that: in the x-direction, it's $(x - a)$ away, and in the y-direction, it's $(y - b)$ away. That means

$$(x-a)^2 + (y-b)^2 = r^2$$

That's the equation of the circle with a radius of r centred at (a, b) – a very useful thing to remember. Figure 11-1 shows this derivation!

Rearranging circle equations

Most of the topics at A level have what I would call a 'classical' question – the kind of question they often fall back on and that you'd be foolish to go into an exam not knowing how to answer. In the case of circles, the question is something like this:

The equation of a circle is $x^2 + y^2 + 6x - 8y = 0$. Find the radius and the coordinates of the centre of the circle.

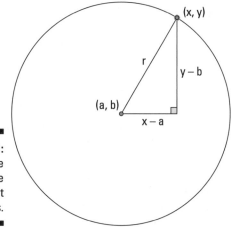

Figure 11-1:
Why a circle
has the
equation it
does.

I think that calls for a recipe, don't you?

1. **Write out the equation of a circle.**

   $$(x-a)^2 + (y-b)^2 = r^2$$

2. **Expand the brackets.**

   $$x^2 - 2ax + a^2 + y^2 - 2by + b^2 = r^2$$

3. **Work out what a and b are by comparing the coefficients of x and y to those in the equation you're given.**

   You need to match $x^2 - 2ax + a^2 + y^2 - 2by + b^2 = r^2$ with $x^2 + y^2 + 6x - 8y = 0$. Looking at the number of xs and number of ys in each equation, you see that $a = -3$ and $b = 4$, meaning that the centre is at $(-3, 4)$.

4. **Now replace a and b with the numbers you worked out in Step 3 and work out an equation with just r in it. The trick is to make the left-hand side of both equations the same.**

   Your equation $x^2 - 2ax + a^2 + y^2 - 2by + b^2 = r^2$ can be rearranged to $x^2 + y^2 + 6x - 8y = r^2 - 25$. You know from the circle equation that $x^2 + y^2 + 6x - 8y = 0$, so $0 = r^2 - 25$.

5. **Solve this for r, taking only the positive root (because a distance must be positive).**

   Here, $r = 5$.

6. **Write down your answer clearly.**

   The centre is at $(-3, 4)$, and the radius is 5.

Solving circle equations

The equation of a circle works exactly like any other equation. Here are some typical things you may be asked to do:

- **Find where the circle crosses the *x*- or *y*-axis.** Here, all you do is substitute $y = 0$ or $x = 0$ into the equation and solve the resulting quadratic.

- **Find where the circle crosses some other line or curve.** In this case, you have nonlinear simultaneous equations. Just apply the techniques from Chapter 3 to the two equations.

- **Work out whether a line intersects (or touches) the circle.** This type of problem is similar to the preceding one, although after you've formed an equation in one variable, you should *count* the solutions instead of finding them. Use the discriminant of the quadratic, as in Chapter 6.

In all cases, when you've finally solved the equations (for *x* and *y*, if necessary), give your answer or answers in the form you've been asked for: if they want the coordinates of a point, give them the coordinates of a point. If they want an *x*-value, give them an *x*-value. Most mark schemes make some allowances for giving answers in the wrong form; in some cases, it will lose you a mark or two. In all cases, it will annoy whoever's marking your paper. Make that person's life easy!

More common than these questions, though, are questions that explicitly involve tangents. Read on!

Tackling tangents

A *tangent* to a circle is any line that just grazes the curve. A *chord* is a line segment between any two points on the circle. Both are shown in Figure 11-2.

Figure 11-2:
A radius
(*OC*), a tangent (*CT*)
and a chord
(*AB*). The
tangent is at
right angles
to the
radius.

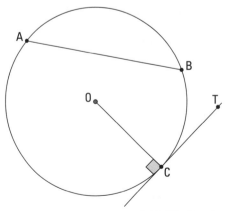

© John Wiley & Sons, Inc.

In this section, you make use of two circle theorems (you didn't believe your teacher when he said they'd be useful, did you?) to find the equation of a tangent or the length of a tangent. There aren't any sums involving chords in this section, but it's as well to know what a chord is.

Finding the equation of a tangent

Here's the circle theorem you need: a tangent to a circle is perpendicular to the radius, as shown in Figure 11-2. That makes finding the equation of a tangent to a circle through a given point a fairly simple process: you need to find the equation of the line through that point, with a gradient you can work out. In this example, the circle has its centre at $(-3, 4)$ and a radius of 5; suppose you want the tangent at $(1, 7)$ in the form $ax + by + c = 0$.

1. **Work out the gradient of the radius.**

   The change in x is the difference between the x-coordinates of the point and the circle's centre; the change in y is the difference in their y-coordinates. That gives you a gradient of $\frac{7-4}{1-(-3)} = \frac{3}{4}$.

2. **Find the gradient of the tangent.**

   The gradient of the tangent is the negative reciprocal of the radius gradient, because the tangent is at right angles to the radius. You have $m = -\frac{4}{3}$.

3. **Use the straight line equation.**

   Put the coordinates and the gradient into the equation $(y - y_0) = m(x - x_0)$ and rearrange as needed:

   $$(y - 7) = -\frac{4}{3}(x - 1)$$
   $$3y - 21 = -4x + 4$$
   $$4x + 3y - 25 = 0$$

It's always worth checking that your answer a) makes sense (does the point you used to create the line actually lie on it?) and b) is in the form they asked for. If they'd asked for $y = mx + c$, you would rearrange accordingly.

Finding the lengths of tangents (and related things)

The relevant circle theorems here are a) the one from the preceding section (a tangent is at right angles to the appropriate radius) and b) that tangents extending to the same point outside the circle have the same length.

The range of things you may be asked about tangents is quite wide and varied, so here are some general tips for approaching the kind of question that requires the length of a tangent:

- ✓ **Draw a big picture.** Being able to see what's going on *really* helps you figure out what shapes are what.

- ✓ **Look for triangles.** Triangles you can solve, especially right-angled triangles (typically involving a tangent and a radius) and isosceles triangles (two radii or two tangents), are shapes you've been playing with forever.

- ✓ **Think about what would be useful to know.** Almost always, a missing triangle side or angle is something you can work out from what you've been given.

For example, sticking with the circle centred at O (–3, 4) with radius 5, you might be told that a tangent passes through the point P (12, 3) and have to find the exact length of the tangent.

If you draw a sketch like the one in Figure 11-3a and think about what you know about the circle, you may draw a radius to either of the points where the circle meets the tangent, at T and T'. You now have the two short sides of a right-angled triangle, the hypotenuse of which connects the centre O (at (–3, 4)) to the point P (at (12, 3)). Using Pythagoras's theorem, the length of $\overline{OP}$ is $\sqrt{226}$. Now you've reduced the problem to finding a short side of a right-angled triangle, knowing the other two; you can use Pythagoras again to work out that TP is $\sqrt{201}$.

Inside/outside

Deciding whether a point lies inside or outside a circle is something the exam creators ask occasionally, just to keep you on your toes. It's an easy question to answer if you think of where the equation of a circle comes from. Here it is again:

$$(x-a)^2 + (y-b)^2 = r^2$$

It means 'The distance of (x, y) from (a, b) is r.' For points inside the circle, the distance is less than r, so if you substitute the coordinates of the point you're interested in into $(x-a)^2 + (y-b)^2$ and get a value *smaller* than r^2, you're inside the circle. Similarly, if it's bigger than r^2, you're outside the circle. Simple!

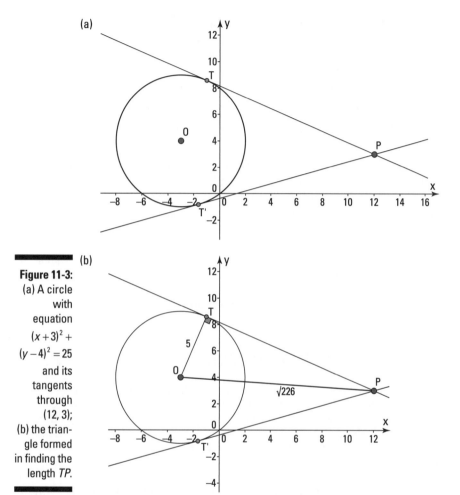

Figure 11-3:
(a) A circle
with
equation
$(x+3)^2 +$
$(y-4)^2 = 25$
and its
tangents
through
(12, 3);
(b) the trian-
gle formed
in finding the
length *TP*.

Rocking Out with Radians

If I could change one thing about the modern world, it would be this: I would abolish degrees. Get rid of them completely. Terrible, stupid, horrible way of measuring angles. What does 360 have to do with a circle? Nothing, absolutely nothing. The Sumerians who came up with the blithering idea presumably thought that's roughly how many days there are in a year and that 60 was a pretty cool number. They might as well have plucked 30 or 100 or any other number out of the air: it's nothing inherently to do with circles.

Radians, on the other hand, are lovely. Glorious, beautiful radians: 2π of them to the circle, which makes sense: the circumference of a circle is $2\pi r$. Among the reasons radians are way better than degrees:

- ✔ 2π is something to do with a circle; 360 isn't.
- ✔ The formulas for arc length and sector area are *much* simpler in radians.
- ✔ For small angles, $\sin(x) \approx x$, $\tan(x) \approx x$ and $\cos(x) \approx 1 - \frac{1}{2}x^2$ – but only if you measure things in radians.
- ✔ If you want to do calculus with trigonometric functions (and if you're going to do A2, it doesn't matter whether you *want* to, you will most certainly have to), working with radians avoids all manner of scale-factor issues.
- ✔ If you want to work with complex numbers (and if you're going to do a maths degree, you most certainly will), you also need to use radians.

In short, you should take your calculator right now and put it into radians mode. Although some questions do ask you to work in degrees, it's easy enough to convert radians into degrees and vice versa – the vice versa, of course, is the important direction to know about.

Converting between radians and degrees

For all the time you spent at GCSE learning to convert between ounces and grams, miles and kilometres, pounds sterling and euros, and anything else you might care to measure in two different units, there's remarkably little conversion involved at A level. The only time I can think of a genuine need to convert anything is if you have an angle in degrees that would be more sensibly expressed in radians.

Rather than telling you what the conversion factors are and expecting you to remember them, I show you how to work out the relationship between them. Remember, there are 2π radians in a whole circle, the same as 360 of those rancid degrees. If you have an angle, you can express that as a fraction of a circle:

$$\frac{\text{angle in degrees}}{360} = \frac{\text{angle in radians}}{2\pi}$$

That means that if you want to convert, say, 72° into radians, you could write down $\frac{72}{360} = \frac{\theta}{2\pi}$, simplify the left-hand side to $\frac{1}{5}$ and multiply everything by 2π to get $\theta = \frac{2}{5}\pi$.

Similarly, if (for some unearthly reason) you wanted to convert $\frac{5}{6}\pi$ radians into degrees, you could set up the fractions as $\frac{x}{360} = \frac{\left(\frac{5}{6}\pi\right)}{2\pi}$. Simplifying the right-hand side gives you $\frac{5}{12}$, and multiplying everything by 360° gives you $x = 150°$.

REMEMBER

It's *very* unusual for an angle to be more useful in degrees than in radians. If you're not explicitly told to use degrees in a question, you should assume that radians are the units to use. Also, if you're writing an angle in degrees, you ought to use the ° symbol.

Finding arc lengths

For a simple demonstration of the inherent superiority of radians (What's that? Yes, I know I'm banging on about it. It's important!), you need look no further than the formula for the length of the arc of a circle. You can see an arc (and a sector) in Figure 11-4.

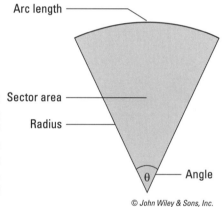

Figure 11-4:
A sector of
a circle, its
arc length
and its area.

Arc length

Sector area

Radius

θ — Angle

© John Wiley & Sons, Inc.

If you wanted to work out the arc length (usually a curly *l* that looks like this: ℓ) given the circle's central angle in degrees – let's say $x°$ – and the circle's radius (r), you would work out the fraction of the circle you have $\left(\frac{x}{360}\right)$, find the circumference of the circle ($2\pi r$), and multiply them together to get something like $\ell = \frac{\pi x r}{180}$. Any chance of remembering that? Thought not.

If you have the angle (θ) in radians instead, as nature intended, the fraction of the circle is $\frac{\theta}{2\pi}$, and multiplying that by the circumference gives you $\ell = r\theta$. Now *that's* a formula that isn't going to take too much memorising.

TECHNICAL STUFF

A *radian* is defined as the angle at the point of a circular sector such that the arc length is the same as the radius. It's no coincidence that *radius* and *radian* are such similar words!

Finding sector areas

You can find the area of a sector (see Figure 11-4) by working out what fraction of a circle you have and multiplying that by the circle's area. Or you can remember a simple radian formula – it's up to you.

If you have an angle in degrees ($x°$) and a radius, your sector constitutes $\frac{x}{360}$ of the circle. The area of the whole circle is πr^2, so your sector's area is $\frac{\pi r^2 x}{360}$.

If you do that in radians instead, with an angle of θ, you have $\frac{\pi r^2 \theta}{2\pi} = \frac{1}{2} r^2 \theta$. That's a much simpler thing to remember – especially after you spot that it's the integral of the arc length formula (with respect to r).

A Core 2 question may give you a sector like this one, with a radius of 8 cm and an angle of $\frac{4}{7}\pi$ at the centre. Suppose the question asks for the area of the sector and its perimeter.

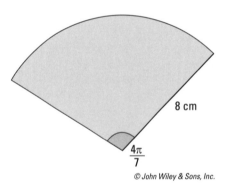

8 cm

$\frac{4\pi}{7}$

© John Wiley & Sons, Inc.

For the area, substitute your values into the formula $A = \frac{1}{2}r^2\theta$ and work out the answer (leave it as a fraction, in terms of π, unless you're explicitly told otherwise):

$$A = \frac{1}{2} \times 64 \times \frac{4}{7}\pi$$
$$= \frac{128}{7}\pi$$

For the perimeter, you already know two sides; they're both 8 cm. All you need is the remaining side. Substitute the values of r and θ into the $\ell = r\theta$ equation to get $\ell = 8 \times \frac{4}{7}\pi = \frac{32}{7}\pi$. Now combine the three sides into a single answer: $16 + \frac{32}{7}\pi$ is probably as nice as it gets.

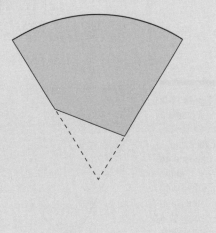

Negative space

My little brother, who is a graphic designer, occasionally mutters something from behind his stubble, horn-rimmed glasses and turtle-neck jumper about 'negative space'. I don't know what that means in his world, but in maths, it means looking at a shape like the following one and saying, 'The shaded area must be the area of the sector minus the area of the triangle.'

It's quite a key skill – I mention it in Chapter 16, on finding areas by integration, as well – and it's a common enough stumbling-block that I'd say: if you're stuck on this kind of question, try looking for a shape that's been taken away and see if you can add it back on.

Taking Care of Triangles and Segments

Almost every Core 2 paper has a nasty question involving an emblem or a garden (or some other pretend bit of context) in the shape of a triangle with a segment of a circle added to it, or a triangle with a sector cut out, or any number of variations on the same theme.

Here is where your problem-solving skills are tested: in all these examples, the trick is to break the shape down into smaller shapes you know how to deal with – typically, triangles, sectors and segments.

Solving triangles

You actually know all the triangle techniques you could possibly need (see Chapter 4), but here's a good place for a recap of the rules – Figure 11-5 shows how the triangle is labelled (all these rules work just as well in radians as in degrees):

- ✔ The angles in a triangle add up to 180° or, if you're a grown-up, π radians.
- ✔ The area of a triangle is $\frac{1}{2}bh$, where b is the base and h the perpendicular height.
- ✔ Alternatively, the area is $\frac{1}{2}ab\sin(C)$.

✔ If you know a side and its opposite angle and anything else at all about the triangle, you may use the sine rule to find the remaining angles and sides: $\dfrac{a}{\sin(A)} = \dfrac{b}{\sin(B)} = \dfrac{c}{\sin(C)}$.

✔ If none of the above works and you still need a missing side or angle, try the cosine rule: $a^2 = b^2 + c^2 - 2bc\cos(A)$.

Figure 11-5:
A labelled triangle. Every side is opposite an angle marked with the same letter (but capitalised).

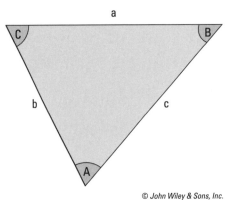

© John Wiley & Sons, Inc.

Sorting out segments

A *segment* is the part of a circle cut off by a chord, as in Figure 11-6. You can – and should – think of it as a sector of a circle with a triangle removed; when you have that in mind, how to figure out its area becomes almost obvious.

Figure 11-6:
A segment of a circle.

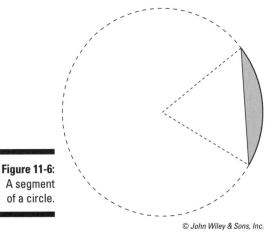

© John Wiley & Sons, Inc.

To find the area of a segment of a circle, here's what to do:

1. **Write down the radius of the circle.**

   This is usually given to you.

2. **Work out the angle at the centre of the circle.**

   This may require a bit of cosine rule; alternatively, it may be given to you.

3. **Work out the area of the sector using the $A = \frac{1}{2}r^2\theta$ formula.**

4. **Work out the area of the triangle using the $A = \frac{1}{2}r^2\sin(\theta)$ formula.**

5. **Subtract the two to get the sector area.**

For example, if your circle has a radius of 6 cm and the angle at the centre is $\frac{1}{6}\pi$, the sector area is 3π and the triangle area is 9; the area of the segment is therefore $(3\pi - 9)$ cm^2.

Putting it all together

Each of the individual parts of a complicated sectors and segments and triangles question is fairly straightforward. For most students, the problem comes in putting everything together.

As with any complicated question, there are two things that will help you enormously: firstly, draw a big picture with all the information in it; and secondly, when you write down something you've worked out, be sure to write down what it is, too! I mean, don't just write '0.7 radians'; write 'Angle $BCD = 0.7$ radians'. It saves you from having to work things out over and over again.

A question from the complex end of the exam might give you a diagram like the following one and tell you this:

> *ABE* and *CBD* are congruent triangles, and *ABC* is a straight line. The lengths of $\overline{AB}$ and $\overline{EA}$ are 10.5 and 8 cm, respectively. The angle *EAB* is 0.65 radians. The points *D* and *E* are joined by a straight line and by an arc of a circle with centre *B*. Find the area of the segment formed by this line and this arc.

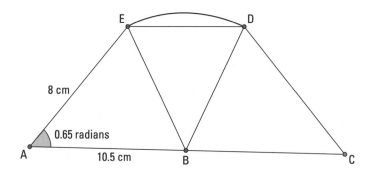

This question is a real test of your problem-solving skills – and you could expect at least eight marks for it in an exam. Here's how I'd approach it, which might seem backwards at first. Hear me out!

1. **Ask: what would make this question easy, and how can I get that information?**

   To find the area of a segment, you need the radius and the angle at the centre of the circle. You can get the radius of the circle by finding the missing side of triangle *ABE;* you can get the angle at the centre by finding the angles on either side of it.

2. **Find the radius.**

   You have an angle and the sides beside it, so use the cosine rule:

   $$BE^2 = 10.5^2 + 8^2 - 2(10.5)(8)\cos(0.65)$$
   $$= 174.25 - 168\cos(0.65)$$
   $$\approx 40.51 \text{ cm}$$

   Therefore, the radius is the square root of that, 6.365 cm.

3. **Find the angle.**

   You can find angle *EBA* using the sine rule and your answer from Step 2:

   $$\frac{6.365}{\sin(0.65)} = \frac{8}{\sin(EBA)}$$
   $$\sin(EBA) = \frac{8\sin(0.65)}{6.365}$$

   Using your 'Ans' button rather than the rounded version, you get $\sin(EBA) = 0.761$ and $EBA = 0.864$ radians, approximately. Because triangle *ABE* is congruent to triangle *CBD*, angle *DBC* is also 0.864 radians. Because *ABC* is a straight line, the three angles at *B* add up to π, so angle *EBD* is $\pi - 2 \times 0.864$, which is 1.413 radians.

4. Find the segment area.

Remembering that the formula is 'sector minus triangle' and calling angle *EBD* θ, you can write down $A = \frac{1}{2}r^2\theta - \frac{1}{2}r^2\sin(\theta)$. If you factorise that to get $A = \frac{1}{2}r^2(\theta - \sin(\theta))$ and fill in the gaps, you get

$$A \approx \frac{1}{2} \times 40.51 \times (1.413 - 0.988) = 8.614 \text{ cm}^2$$

Wherever possible, instead of typing out the decimals each time, use your calculator's 'Ans' button to store answers, as I've done here (while reporting my numbers to four significant figures). As a rule of thumb, you'll probably be OK if you use one more significant figure than the question asks for, but a) it's much less typing to press 'Ans', and b) why would you want to round if you didn't need to?

This is a brute of a question – in an exam, you could reasonably expect some hints to help you through it. This particular example would most likely be split into three parts: the first asking you to find the length *BE*, the second asking you to show that the angle *EBD* is 1.413 radians, and the last part asking for the segment area.

Chapter 12

Taking Trigonometry Further

● ●

In This Chapter

▶ Sketching basic trig functions

▶ Using trig identities and compound angle formulas

▶ Finding multiple solutions

▶ Completing trig proofs

● ●

*A*s you might expect from a maths qualification designed to take your understanding further, the trigonometry you did at GCSE gets a lot more involved at A level. In some respects, it gets simpler, as you begin to see how everything links together, but in most respects, it's messy and a little confusing. (You can handle it, though. I believe in you!)

In A level trigonometry, my advice is that radians are the *correct* way to measure angles, unless you're explicitly told otherwise (for example, if the question asks for an angle to the nearest degree or tells you $0 \le x \le 180°$). This chapter uses radians almost exclusively (the full rant explaining why is in Chapter 11).

Here, I show you how to exploit symmetry to draw awesome graphs of the trig functions and how to use right-angled triangles to remember your identities. You get to grips with compound angles and finding all the solutions to trigonometric equations, and you discover how to prove things on demand.

Sketching Up Symmetries

A little secret: I *love* the basic trig functions. I love the way you can reflect and rotate them without changing their nature. I love the way that differentiating sine and cosine gives you variations on sine and cosine. I love that you can express them in terms of *e*, which you won't see unless you do Further Maths and/or university maths. I love the way that adding them together gives you another variation on sine and cosine. I love their link with Pythagoras. There's just the right balance of change and constancy for my taste, so I award $\sin(x)$ a Lifetime Achievement Award for being a brilliant function, and $\cos(x)$ is in the running for Best Supporting Role.

The symmetries are the best bit, though. Sketching the graphs of the trigonometric functions is by far the most obvious way to see which symmetries you can use. These little drawings will come in very useful in later sections!

There's one big rule for sketching all six of the basic trigonometric graphs (by which I mean, not scaled or translated): interesting things happen every quarter-circle. By 'an interesting thing', I mean one of the following happens:

- ✔ The graph crosses the x-axis.
- ✔ The graph reaches a minimum or maximum.
- ✔ The graph has a vertical asymptote.

And here are the three related symmetry rules:

- ✔ The graph has reflective symmetry about the normal at a turning point.
- ✔ The graph has rotational symmetry about a point where the graph crosses the x-axis.
- ✔ The graph has rotational symmetry about a point where the graph's vertical asymptote crosses the x-axis.

You can check back to Chapter 10 to see how these rules apply to the curves of the sine, cosine and tangent functions.

Identifying Trig Identities

You're hopefully aware of five ways of relating two expressions already: they can be equal (=), or the first can be larger than (>), smaller than (<), at least as much as (≥) or no more than (≤) the other. In this section, I introduce another relationship: *equivalent to* (≡), an equals sign with three bars. (It's also called *identically equal to*, but I'm too lazy to type that out repeatedly.)

You use an 'equivalent to' sign to show that something is *always* true, no matter which value of the variable you use. For example, strictly speaking, $(x+2)^2 \equiv x^2 + 4x + 4$ – you can use any value of x, because you get the same thing on both sides.

In your A level, the 'equivalent to' sign crops up most commonly in trigonometric identities, of which there are two must-knows and three further nice-to-know-but-easy-enough-to-work-outs.

Relating trig functions with the basic triangle

Imagine a right-angled triangle where one of the other angles is θ and the hypotenuse has length 1. Or, if you prefer, you can look at Figure 12-1, where I've drawn it for you.

A little bit of SOH CAH TOA tells you the adjacent side has length $\cos(\theta)$. The opposite side, similarly, has length $\sin(\theta)$. You can use this to generate the two must-know trig identities:

> ✔ $\tan(\theta) \equiv \dfrac{\sin(\theta)}{\cos(\theta)}$, because it's opposite divided by adjacent
>
> ✔ $\sin^2(\theta) + \cos^2(\theta) \equiv 1$, because Pythagoras says so

These two identities, plus the definitions of the minor trig functions, are all you need to generate the other three trig identities, which come up in the next section.

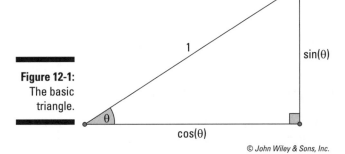

Figure 12-1: The basic triangle.

© John Wiley & Sons, Inc.

Relating the minor trig functions

It turns out that you can draw similar triangles for the minor trig functions (see Figure 12-2). In the triangle on the left, I've taken the adjacent side to have length 1. That means the opposite side is $\tan(\theta)$ and – with a little work – the hypotenuse is $\dfrac{1}{\cos(\theta)}$, or $\sec(\theta)$. That means

$$\sec^2(\theta) \equiv 1 + \tan^2(\theta)$$

On the right, the opposite side of the triangle has length 1. That makes the adjacent side $\dfrac{1}{\tan(\theta)}$, or $\cot(\theta)$; meanwhile, the hypotenuse is $\dfrac{1}{\sin(\theta)}$, or $\operatorname{cosec}(\theta)$. That gives you the other identity you're interested in:

$$\operatorname{cosec}^2(\theta) \equiv 1 + \cot^2(\theta)$$

Another way to get these identities is to divide $\sin^2(\theta) + \cos^2(\theta) \equiv 1$ by $\cos^2(\theta)$ or $\sin^2(\theta)$.

(a) 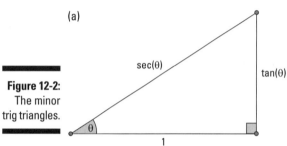 (b)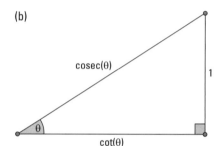

Figure 12-2:
The minor trig triangles.

© John Wiley & Sons, Inc.

The fifth identity I mentioned? That one doesn't have a triangle attached to it, really: it's simply the reciprocal of $\tan\theta \equiv \dfrac{\sin\theta}{\cos\theta}$, which is $\cot\theta \equiv \dfrac{\cos\theta}{\sin\theta}$.

Relating sine, cosine and tangent

A common Core 2 question gives you a trig function and asks for another trig function using the same angle. For example, the question may give you $\tan(\alpha) = \dfrac{3}{5}$, tell you that α is a reflex angle and ask for $\cos(\alpha)$. Here's how you tackle that sort of question – and, as a special treat, I'll use degrees – and show you two ways!

1. **Find a possible answer for α by typing $\tan^{-1}\left(\dfrac{3}{5}\right)$ into your calculator.**

   You get approximately 31.0°. Unfortunately, that's not a reflex angle.

   Reflex angles are between 180° and 360°, as opposed to acute angles (0° to 90°) and obtuse angles (between 90° and 180°).

2. **Sketch the graph of $y = \tan(x)$ between 0° and 360°, and draw a horizontal line where y equals about $\dfrac{3}{5}$ (it doesn't have to be exact; it's just a visual aid).**

   The horizontal line should cross the tangent graph twice (once, hopefully, near 30°).

3. Use the symmetries of the graph to find the reflex solution.

The first solution is about 31.0° to the right of where the graph crosses the x-axis, so the second solution is about 31.0° to the right of the *other* place where the graph crosses the x-axis, at 180°; therefore, the solution is near 211.0°.

4. Now work out the cosine of this angle.

$$\cos(211°) \approx -0.857$$

If possible, use your calculator's 'Ans' button rather than typing in 211.0 over and over again. The value isn't precisely 211.0 or even 210.963757 – the number on your calculator is a rounded version.

This method is OK, but it's not good if you've been asked for the *exact* value of the cosine, which is the more normal thing to ask. In that case, you're going to need a triangle!

1. Draw a triangle that has an angle with a tangent of $\frac{3}{5}$, like the one in Figure 12-3a.

2. Work out the length of the remaining side using Pythagoras.

The hypotenuse is $\sqrt{34}$.

3. Now work out the cosine, which is $\frac{5}{\sqrt{34}}$.

4. Last thing, work out which sign the cosine ought to have. Sketch the tangent and cosine graphs (really quickly!) and the line $y = \frac{3}{5}$, as in Figure 12-3b.

The only places the line crosses the $y = \tan(x)$ curve are between 0° and 90° and between 180° and 270°. You know α is reflex (between 180° and 360°), so α has to be the value that's between 180° and 270°. There, the value of cosine is definitely negative, which means $\cos(\alpha) = -\frac{5}{\sqrt{34}}$.

Reassuringly, that works out to –0.857, the same answer as before!

Having both of these methods at your disposal gives you a good check on your answer!

There's an ambiguity in the sign because when you're doing trigonometry all the way around a circle, distances (or rather, displacements) can be negative, so you need to take that into consideration! In Figure 12-4, you can see the idea of negative displacements; the sides of the triangle are actually –5 (moving left), –3 (going down) and $\sqrt{34}$.

(a)

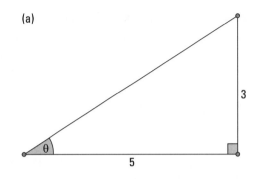

(b)

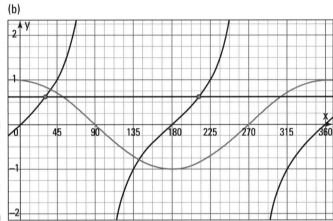

Figure 12-3:
A triangle
and
necessary
graphs.

© John Wiley & Sons, Inc.

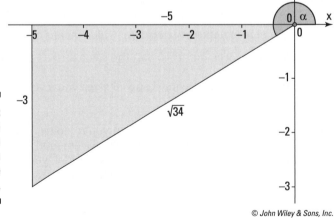

Figure 12-4:
A triangle
with side
lengths in
the negative
direction.

© John Wiley & Sons, Inc.

Going around in circles

'What does sine *actually mean*?' is one of the most common questions I'm asked in trigonometry, and 'the ratio between the lengths of the side opposite your angle and the hypotenuse' is a bit unsatisfactory, especially when you're talking about $\sin(230°)$, an angle that doesn't fit in a triangle. Luckily, there's another explanation.

Take a circle with a radius of 1 unit, like the one in the next diagram. Imagine you're facing east (along the positive *x*-axis). Turn anticlockwise through the angle you're interested in (so, if your angle is $\frac{1}{4}\pi$, you're facing north-east; if it's $\frac{3}{2}\pi$, you're facing south) and walk to the edge of the circle. The sine of your angle is the *y*-coordinate you end up at; the cosine of the angle is the *x*-coordinate. The tangent, meanwhile, is the gradient of the line you walked along.

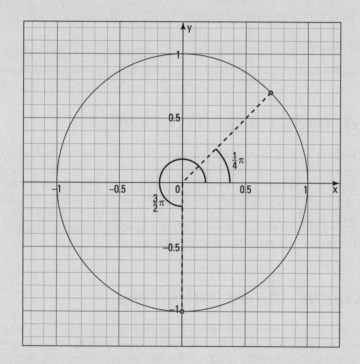

This explains an awful lot about the graphs!

✔ The graph of $y = \sin(\theta)$ starts at (0, 0) and rises quite steeply before tapering off and hitting a peak at $\left(\frac{1}{2}\pi, 1\right)$, heading north on your circle. It then drops, slowly at first but then more quickly, towards $(\pi, 0)$ in the west, and it carries on down to $\left(\frac{3}{2}\pi, -1\right)$ in the south before rising back up to $(2\pi, 0)$ in the east.

✔ The graph of $y = \cos(\theta)$ starts at (0, 1) in the east and drops, slowly at first but then more quickly, towards $\left(\frac{1}{2}\pi, 0\right)$

(continued)

(continued)

in the north. It carries on down to $(\pi, -1)$ in the west before rising back up to $\left(\dfrac{3}{2}\pi, 0\right)$ in the south. It finally crests again at $(2\pi, 1)$ in the east.

✔ The graph of $y = \tan(\theta)$ needs a bit more thought. The gradient is 0 when the angle is 0, but it gets steeper as θ increases. At $\theta = \dfrac{1}{4}\pi$, heading north-east, the gradient is 1, but at $\theta = \dfrac{1}{2}\pi$, when you're facing north, the line's gradient is undefined. (Slightly east of north, it's a huge positive number; slightly west of north, it's a steep line with a negative gradient.) The negative gradient gets closer to 0 as the angle increases further: heading north-west at $\theta = \dfrac{3}{4}\pi$, the gradient is –1; in the west at $\theta = \pi$, the gradient is 0. The pattern then repeats.

✔ The circle also explains why all the graphs go on forever in both directions. If you turn $\dfrac{1}{4}\pi$ radians *clockwise* (so $\theta = -\dfrac{1}{4}\pi$), you're facing the south-east, as if you'd turned $\dfrac{7}{4}\pi$ radians anticlockwise, or even $\dfrac{15}{4}\pi$ – turning an extra 2π radians either way doesn't change your coordinates!

Taming Trigonometric Proofs

The vast bulk of the proofs you see at A level involves showing that one expression is exactly the same as another. If you were an engineer or an economist (shudder), you might say, 'I know! I'll check the expressions with a few numbers!' and when they all came out right, you'd say, 'Look, they're all the same – QED.' Luckily, you're a mathematician, and you have higher standards than that. You need to use the rules of maths to show that you can manipulate one expression into the other.

Laying things out nicely

In GCSE English, did you have a rough book and a neat book so you could draft an essay without worrying about presentation before you submitted your final version? That's not a bad idea. Proofs really have as much in common with essays as they do with sums: you're trying to convince someone (your teacher or an examiner, usually) that something is true. Doing a rough draft – or at least some scratch working – elsewhere can help make your argument more solid.

In maths, unlike in English, there's no room for rhetoric or bluffing – you're much more limited in how your argument can progress (even though it often seems like there are infinitely many ways you could go).

Here are a few pointers about how I like to see proofs laid out for clarity:

- ✔ Start by writing out what you're trying to prove.

- ✔ Work on one side at a time.

- ✔ Your goal (usually) is to show that one side is equivalent to the other. Start with LHS = . . . (meaning left-hand-side) and work with it, on its own, until you get to = RHS (for right-hand-side), or vice versa.

- ✔ If you're not sure where you're going, try working on the proof off to one side or on a different page. Nothing makes a proof more confusing than lots of abandoned starts and crossed-out working.

- ✔ When you're done, either write 'QED' or 'as required' or (my favourite) draw a *Halmos*: ∎.

There's a critical difference between developing a proof and solving an equation: in a proof, you can't assume the two expressions are the same (until you've finished). That means you're not allowed to do the same thing to both expressions; you have to work with each expression on its own, finding equivalent expressions until you reach your target.

Perfecting your proof techniques

Almost all the proofs you see at A level are of the 'direct' sort: you are given one expression (usually a horrible mess) and are asked to prove or show that it's the same as something else (usually much nicer). Once in a while, you may be asked to show that something isn't true, which you can do by finding a *counterexample* – for example, if I say, 'all swans are white', you can prove me wrong by showing me a black swan. Someone should write a book about that.

Similarly, if you're asked to show that the statement '$\cos(x) + \sin(x) < \sqrt{2}$ for all values of x' is untrue, you just need to find a value of x for which the statement isn't true. You could say, 'If $x = \frac{1}{4}\pi$, then $\cos(x) + \sin(x) = \sqrt{2}$, which disproves the claim.'

A *direct* proof usually consists of showing that two expressions are equivalent by manipulating either or both of them into the same form. You can see one of these proofs in the later section 'The formulas that don't come up but I have to mention anyway'.

Obviously, I can't tell you exactly what to do in every case – there are infinitely many variations. However, some techniques are especially useful:

- ✔ **Combine fractions.** Turning a pair of fractions into a single one is almost always a good step. Be sure to use your fractions rules from Chapter 3; they apply to fractions with trigonometric stuff as much as they do anywhere else.

- ✔ **Play 'spot the difference'.** Seeing what you need and what you need to get rid of to match the other side can make it easier to figure out which identities to use. If you've got a $\sin^2(x)$ loitering about and you're missing a $\cos^2(x)$, there's a good chance you need to convert the one into the other.

- ✔ **Factorise and simplify.** If you can take a factor out, do. If you can simplify a fraction, do.

- ✔ **Watch out for the arguments.** I often used to overlook the bits in the brackets and would mistake $\sin(2x)$ for $\sin(x)$ or similar. Be careful not to muck them up!

- ✔ **Don't be afraid to try things!** If you've got funny things in the brackets (like 40° and 50°, for instance), try replacing 50° with $(40° + 10°)$ – which doesn't work out nicely, too bad – or with $(90° - 40°)$, which does.

Finding mistakes

I've seen it a hundred times: a student attacks a proof, makes good progress and then suddenly slumps. 'I'm stuck,' the student says, or 'I don't know where I'm going' or 'Is this right?' There's a way to tell whether you're on track, though, and it's a trick I learnt from an engineer: try a few values.

Stick the same random value for x (or θ or whatever variable you're working with) into each of your expressions. If they're the same, you have some evidence that you're on the right track. But note that this approach isn't foolproof. Even if you try several values and they all come out right, all you have is evidence, which isn't the same thing as proof. However, it should be enough to convince you practically, even without being mathematically certain.

On the other hand, if your answers are different, then you've *definitely* made a mistake somewhere. Try the same trick a bit further back in your working, and narrow down where your mistake is. Put it right. Carry on.

Clearing Up Compound Angles

Be really careful: it's incredibly easy to see something like $\cos(A + B)$ and mistakenly believe you can expand it as $\cos(A) + \cos(B)$. You can't, except for special values of A and B (which happen to make a very nice graph, if you're at a loose end and fancy firing up a graphing program like desmos.com).

The formulas for expanding *compound angles* (literally, angles made up out of other angles) are exactly where you'd expect to find them: in the formula book. However, they're worth knowing, and here's why: if you copy stuff down blindly without ever learning it, you never develop a sense for what looks right and what smells funny. You never get the sixth sense that says,

'Oh, I can combine those' or 'Hang on, that minus sign is out of place – better check.' Lecture over.

In this section, I go through the compound angle formulas as they're given, look at some special cases that come up frequently (such as doubling or tripling the angle), and take you through the $R\sin(x+\alpha)$-style questions.

Lastly, the formula book lists some compound angle formulas below the main ones. It's *extremely* unusual to see these come up in an exam, but they're technically fair game, and with the recent trend towards deliberately setting bloodbath questions, I'd not be too surprised to see them in the next few years.

Compound angle formulas

Your formula book gives you three compound angle formulas:

> ✔ $\sin(A\pm B)\equiv\sin(A)\cos(B)\pm\cos(A)\sin(B)$
>
> ✔ $\cos(A\pm B)\equiv\cos(A)\cos(B)\mp\sin(A)\sin(B)$
>
> ✔ $\tan(A\pm B)\equiv\dfrac{\tan(A)\pm\tan(B)}{1\mp\tan(A)\tan(B)}$

This may be the first time you've seen the odd-looking $\mp$ sign. It shows up only when you have a $\pm$, and it means 'Whichever sign you picked before, you pick the other one here.' The cosine formula is a shorthand for $\cos(A+B)\equiv\cos(A)\cos(B)-\sin(A)\sin(B)$ and $\cos(A-B)\equiv\cos(A)\cos(B)+\sin(A)\sin(B)$.

If you know the trigonometric ratios of a few common angles, you can use them and the compound angle formulas to work out the ratios of other angles. For example, $\sin\left(\frac{1}{12}\pi\right)\equiv\sin\left(\frac{1}{4}\pi-\frac{1}{6}\pi\right)$. Using $A=\frac{1}{4}\pi$ and $B=\frac{1}{6}\pi$, you get the following:

$$\sin\left(\tfrac{1}{4}\pi-\tfrac{1}{6}\pi\right)\equiv\sin\left(\tfrac{1}{4}\pi\right)\cos\left(\tfrac{1}{6}\pi\right)-\cos\left(\tfrac{1}{4}\pi\right)\sin\left(\tfrac{1}{6}\pi\right)$$
$$=\left(\frac{1}{\sqrt{2}}\right)\left(\frac{\sqrt{3}}{2}\right)-\left(\frac{1}{\sqrt{2}}\right)\left(\frac{1}{2}\right)$$
$$=\frac{\sqrt{3}-1}{2\sqrt{2}}$$

Rationalising, you get $\dfrac{\sqrt{6}-\sqrt{2}}{4}$, which you can check on your calculator.

I'm being a bit obstinate in using radians here – you're slightly more likely to see this in degrees – but the same formulas apply. Radians are just better.

Double-angle formulas

The double-angle formulas are a direct consequence of the compound angle formulas. For example, you could find $\sin(2A)$ by noticing it's $\sin(A+A)$, expanding and simplifying. If you did that, you'd find the following:

✔ $\sin(2A) \equiv 2\sin(A)\cos(A)$

✔ $\cos(2A) \equiv \cos^2(A) - \sin^2(A)$, which can be rewritten as

- $\cos(2A) \equiv 2\cos^2(A) - 1$
- $\cos(2A) \equiv 1 - 2\sin^2(A)$

✔ $\tan(2A) \equiv \dfrac{2\tan(A)}{1 - \tan^2(A)}$

The last of those is significantly less common than the others, but they're all worth knowing. They're especially helpful for three-theta kinds of questions, which might say something like 'Show that $\sin(3\theta) \equiv 3\sin(\theta) - 4\sin^3(\theta)$.' Here's how to do that:

1. **Split up 3θ into 2θ + θ and expand the left-hand side of the given problem.**

   $\sin(2\theta + \theta)$

   $\equiv \sin(2\theta)\cos(\theta) + \cos(2\theta)\sin(\theta)$

2. **Expand everything with a 2θ in it.**

   You're looking for sines, so pick expansions that make your life easy. There's only one expansion for $\sin(2\theta)$, but I'd use $\cos(2\theta) \equiv 1 - 2\sin^2(\theta)$ to make the left-hand side into

   $2\sin(\theta)\cos^2(\theta) + (1 - 2\sin^2(\theta))\sin(\theta)$

3. **Get rid of any rogue things you don't want.**

   For example, that $\cos^2(\theta)$ can become $1 - \sin^2(\theta)$. You now have a slightly messy

   $2\sin(\theta)(1 - \sin^2\theta) + (1 - 2\sin^2(\theta))\sin(\theta)$

4. **Simplify.**

   There's a common factor of $\sin(\theta)$ to both terms, so you get

   $\sin(\theta)\left[(2 - 2\sin^2\theta) + (1 - 2\sin^2(\theta))\right]$

   $= \sin(\theta)\left[3 - 4\sin^2(\theta)\right]$

   $= 3\sin(\theta) - 4\sin^3(\theta)$

   as required.

Even if you'll need to multiply a factor back in later, taking it out of a bracket as soon as you spot it makes the sums you need to deal with simpler.

This is a proof, so you're not allowed to assume the result; you have to start from the left-hand expression and keep generating equivalent expressions until you reach the conclusion.

$R\sin(x+\alpha)$-type questions

Before I start here, I'd like to make one thing absolutely clear: Kate the photographer was framed. The last question of the June 2013 EdExcel Core 3 paper was pretty much the definition of a bloodbath. It concerned a photographer – Kate – taking a picture of a marathon runner – John – and gave a fairly involved blurb about the setup of the road, their speeds and so on. All of it was true, but none of it was especially relevant to the question, which turned out to be a pretty standard compound-angle question for 15 marks; it seems a substantial numbers of students didn't read that far or didn't think carefully about the question, and they blamed poor Kate – a professional just doing her job – for their difficulties. I, for one, would like to apologise to Kate for the abuse she suffered at the hands of the Core 3 community and to use this section to help the community avoid repeating the mistakes it made in 2013.

The archetypal $R\sin(x+\alpha)$ question gives you something like $24\sin(x)+7\cos(x)$ and asks you to express it in a compound angle form – such as $R\sin(x+\alpha)$ or, as in this case, $R\cos(x-\alpha)$. All you need to do is expand the form you've been given and match the coefficients.

Matching coefficients is as simple as following these steps:

1. **Expand your template using the appropriate compound angle formula.**

    In this case, $R\cos(x-\alpha) \equiv R\cos(x)\cos(\alpha)+R\sin(x)\sin(\alpha)$.

2. **Play 'spot the difference' with the expression you're trying to match.**

    Here, the given expression is $24\sin(x)+7\cos(x)$. Rearranging the terms in the expanded template gives you

    $$24\sin(x)+7\cos(x) \equiv R\sin(x)\sin(\alpha)+R\cos(x)\cos(\alpha)$$

    If you count $\sin(x)$s, you've got to match 24 on one side with $R\sin(\alpha)$, so $R\sin(\alpha)=24$. Similarly, $R\cos(\alpha)=7$.

3. **Draw a triangle like the one in Figure 12-5.**

    Let the interesting angle be α and the hypotenuse be R. Basic trigonometry tells you the adjacent side is $R\cos(\alpha)$ and the opposite side is $R\sin(\alpha)$, both of whose values you know. Label the adjacent side here with 7 and the opposite with 24.

4. Find α using SOH CAH TOA.

Because $\tan(\alpha)$ is the ratio of the opposite side to the adjacent, you get

$$\frac{R\sin(\alpha)}{R\cos(\alpha)} = \frac{24}{7}$$

$$\tan(\alpha) = \frac{24}{7}$$

$$\alpha \approx 1.287$$

5. Find R using Pythagoras's theorem.

$R^2 = 7^2 + 24^2$, so $R^2 = 625$ and $R = 25$.

6. Write your answer in the form they asked for.

$$24\sin(x) + 7\cos(x) \equiv 25\cos(x - 1.287)$$

Strictly speaking, that answer isn't quite correct, as α is only approximately equal to 1.287. However, it's good enough for A level. Don't tell the maths police.

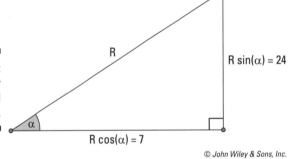

Figure 12-5:
Yet another
helpful
triangle.

R

R sin(α) = 24

R cos(α) = 7

© John Wiley & Sons, Inc.

When life gives you trigonometry, draw triangles.

Expressing expanded trig functions in a compound-angle form isn't just a bit of clever manipulation to pick up three or four marks; it's useful for solving 'real-world' problems too. (By 'real world', I mean the 'real world' examiners live in, where photographers run at precisely 1.68 metres per second.) How is this skill useful? Let me count the ways:

- ✔ **x-values:** If you need to solve for the value of x where the expression takes on a particular value, the question is now effectively a Core 2 problem. You just need to invert the trig function and check you've got the solution or solutions you were asked for.

For example, to find the smallest positive x-value where $24\sin(x)+7\cos(x)=10$, you would use your previous work to turn it into $25\cos(x-1.287)=10$, or $\cos(x-1.287)=0.4$. It's simplest to let $\theta=x-1.287$. Because you want $x>0$, you need $\theta>-1.287$; the principal solution to $\cos(\theta)=0.4$ is $\theta\approx1.159$, but there's also a solution at $\theta\approx-1.159$. That gives you $x\approx0.128$ as the smallest value of x that works.

✔ **Maximum and minimum values:** If you need a maximum or minimum value, think about the trig function: it has a maximum value of 1 and a minimum of –1, so something like $25\cos(x-\alpha)$ has a maximum of 25 and a minimum of –25.

✔ **Locations of the maxima and minima:** If you need to know which value of x gives you a maximum or minimum, just find which values of x give a maximum or minimum in the trig function. (Think: $x=\alpha$ is an interesting angle. Interesting things happen every quarter-circle.)

You might be asked about maxima and minima of something more involved than the original answer – Kate the photographer's speed, for example, worked out to be $\dfrac{21}{25\cos(x-1.287)}$ metres per second – but you can differentiate and solve that without too much trouble. Or, better still, you can spot that the minimum is where the cosine is equal to 1; that's where the argument is 0, so her minimum speed is 0.84 m/s, which happens when $x=1.287$.

The formulas that don't come up but I have to mention anyway

Buried in the formula book beneath the useful compound angle formulas are four further identities:

$$\sin(A)+\sin(B)\equiv2\sin\left(\frac{A+B}{2}\right)\cos\left(\frac{A-B}{2}\right)$$

$$\sin(A)-\sin(B)\equiv2\cos\left(\frac{A+B}{2}\right)\sin\left(\frac{A-B}{2}\right)$$

$$\cos(A)+\cos(B)\equiv2\cos\left(\frac{A+B}{2}\right)\cos\left(\frac{A-B}{2}\right)$$

$$\cos(A)-\cos(B)\equiv-2\sin\left(\frac{A+B}{2}\right)\sin\left(\frac{A-B}{2}\right)$$

The chances of your ever needing to use any one of these, even in an exam, are slim (although, given recent moves towards extra 'rigour' and generally making things harder for the sake of it, you can't rule it out). The most likely thing you'll have to do with these monsters is show that they're true. A question might ask the following:

Use the identities for $\sin(A+B)$ and $\sin(A-B)$ to prove that

$$\sin(P)+\sin(Q) \equiv 2\sin\left(\frac{P+Q}{2}\right)\cos\left(\frac{P-Q}{2}\right).$$

Let's do it!

1. **Decide on how best to manipulate the left-hand expression into the right-hand expression.**

   Taking a hint from the question, let $P = A+B$ and $Q = A-B$.

2. **Apply the identities.**

   $\sin(P) = \sin(A+B)$, which is $\sin(A)\cos(B)+\sin(B)\cos(A)$;

   $\sin(Q) = \sin(A-B)$, which is $\sin(A)\cos(B)-\sin(B)\cos(A)$.

3. **Add these together.**

   You have $\sin(P) + \sin(Q) = \left[\sin(A)\cos(B)+\sin(B)\cos(A)\right]+$ $\left[\sin(A)\cos(B)-\sin(B)\cos(A)\right]$. The $\sin(B)\cos(A)$ terms cancel out to leave you with $\sin(P)+\sin(Q)=2\sin(A)\cos(B)$.

4. **Unfortunately, you made A and B up, so you need to get them in terms of P and Q.**

   Because $P = A+B$ and $Q = A-B$, adding the equations gives you $P+Q = 2A$ and subtracting them gives you $P-Q = 2B$, so you're nearly there!

5. **Tidy up!**

   If $A = \frac{P+Q}{2}$ and $B = \frac{P-Q}{2}$, then the result in Step 3 becomes

   $\sin(P)+\sin(Q) \equiv 2\sin\left(\frac{P+Q}{2}\right)\cos\left(\frac{P-Q}{2}\right)$, as required.

Solutions: Gotta Catch 'Em All

In A level maths, if there's one place where more frustrating marks are dropped than anywhere else, it's in trigonometry questions where several solutions are possible. I've heard many teachers grumble that it's all the fault of the electronic calculator and advocate for banning them – although I bet students made exactly the same mistakes back when you had to read your answers off of tables.

There's a saying in juggling: the problem is that the balls go exactly where you throw them. Similarly, the trouble with calculators is that they answer exactly the question you ask them – and sometimes, you might not realise that there's more to your question than you anticipated.

One of the places that manifests itself is in trigonometry problems. Calculators only ever give you one answer in a trigonometry question – but when you're looking for an angle at A level, more answers are often (in fact, almost always) available.

In fact, unless you put restrictions on your angles (for example, by saying $0 \leq \theta < \pi$), you'll generally get infinitely many solutions. You never have to give infinitely many solutions in an answer, though – at least, not unless you take Further Maths.

Why there are multiple solutions

One reason there are multiple solutions to an equation like $\sin(\theta) = 0.5$ is hidden in the sine graph I've been banging on at you to learn how to sketch all chapter: it goes on forever in both directions, and it's *periodic* – meaning you can horizontally shift the graph a given amount and not realise it's changed. (This is, if you like, another kind of symmetry you can exploit.) In this case, the amount is 2π, which is the same for sine and cosine. The tangent graph, meanwhile, has a period of π.

You can also think of the unit circle I introduced in the 'Going around in circles' sidebar earlier: there are two points on the circle where the y-coordinate is 0.5, and you pass them both every time you go around the circle.

That's not all, though: if you sketch the sine graph (as in Figure 12-6), you'll see that the line $y = 0.5$ cuts the graph of $y = \sin(\theta)$ in more than one place. Your calculator – assuming it's correctly set to radians – will tell you that $\theta = \frac{1}{6}\pi$ is 'the' solution, but as I show you in this section, $\theta = \frac{5}{6}\pi$ is also a perfectly valid angle with a sine of 0.5.

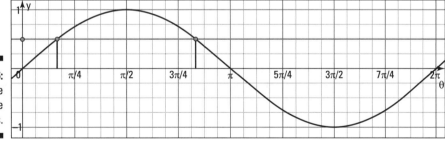

Figure 12-6:
Why there are multiple solutions.

How many answers do you expect?

Working out how many answers you expect to a simple trigonometric problem is dead simple. If you want to know how many solutions there are to $\sin(\theta) = 0.5$ for $0 \le \theta < 2\pi$, here's what you do:

1. **Sketch the graph of $y = \sin(\theta)$ in the given domain.**

   You'll have one complete wave.

2. **Sketch the line $y = 0.5$.**

3. **Count how many times they cross.**

   Hint: it's two, as in Figure 12-6.

Hang on to this sketch. You'll need it for the next section.

TIP

If you have several trigonometric equations to solve at once – for example, if you're dealing with a disguised quadratic (which I cover later in 'Disguised quadratics') – feel free to reuse the same graph for all of them. If you've drawn it big enough, it shouldn't get too cluttered.

Using symmetry to find the answers

After you've drawn your sketch of the trig function (see the preceding section), you can use the symmetry of the graph to find your answers. Here's how to solve $\sin(\theta) = 0.5$ for $0 \le \theta < 2\pi$:

1. **Find one of the solutions using either your calculator or your brain.**

   If you know your triangles, you know that $\sin\left(\frac{1}{6}\pi\right) = 0.5$, and if you don't, your calculator will happily confirm that $\sin^{-1}(0.5) = \frac{1}{6}\pi$.

2. **Find this solution on your graph, and work out how far away it is from an interesting point of your choice.**

   Remember, something interesting happens every quarter-circle. Most simply, you can say the solution is $\frac{1}{6}\pi$ to the right of the crossing point at $(0,0)$, but you may prefer to work out that it's $\frac{1}{3}\pi$ to the left of the maximum at $\left(\frac{1}{2}\pi, 1\right)$.

3. **Relate this to the other point(s) where the line and curve intersect.**

   The second point is $\frac{1}{6}\pi$ to the *left* of the crossing point at $(\pi, 0)$, which means its x-coordinate is $\pi - \frac{1}{6}\pi = \frac{5}{6}\pi$. Alternatively, it's $\frac{1}{3}\pi$ to the right of the maximum at $\left(\frac{1}{2}\pi, 1\right)$, meaning the x-coordinate is at $\frac{1}{2}\pi + \frac{1}{3}\pi = \frac{5}{6}\pi$ Reassuringly, the answer is the same both times.
   Look at Figure 12-7 to see how this works!

Figure 12-7: Using symmetry to find solutions.

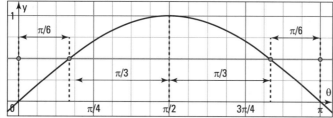

© John Wiley & Sons, Inc.

Checking your answers

The last thing you need to do is to check your answers work. Some things I'd be inclined to check:

- ✔ Do you have as many solutions as you expected? If not, figure out which ones are missing (or which ones don't belong).

- ✔ Are all your solutions in the correct interval? Make sure none are too large or too small.

- ✔ Do all your answers solve the original question? Put the numbers back into the question to make sure.

The Dirty Tricks of Trigonometry

There are (of course) trickier trigonometric problems for you to solve. In order to feel well-and-truly on top of A level trigonometry, you need to be comfortable with solving disguised quadratics (and, once in a while, disguised cubics) and dealing with awkward arguments to the functions.

In this section, I show you how to do both!

Disguised quadratics

There's one sort of trigonometric question you're almost guaranteed to see in Core 2, and you may also come across it in Core 3, so it's important enough to have its own section. It's something that looks like this:

Solve $4\cos^2(x) + 7\sin(x) - 7 = 0$, giving all solutions for $0 \le x < 2\pi$.

Similarly, you may get 'Solve $2\cot^2(\theta) + 3\csc(\theta) - 7 = 0$' in Core 3.

Luckily, there's really only one method for solving this kind of problem, and when you know it, you have an extra half-dozen or so marks nailed on in the exam. Hooray!

Turning a trigonometric mess into a nicer form

Quite often, a disguised quadratic question will give you a hint along the way, saying something like 'Show (this unholy mess) can be written as (a slightly holier mess).' That's a nice kind of question: you can at least see where you're going, and it makes it possible to find the solutions even if you can't work the first part out. However, you don't always get the hint, so make sure you can do these kinds of problems without any extra help.

The trick is to always get everything in terms of a single function – in Core 2, that's bound to be sine or cosine, but Core 3 is open season, and any of the six major or minor trig functions are fair game. How do you get things in terms of a single function? You use your identities.

1. **Find an identity to apply to whichever trig function is squared to make it match the other trig function.**

   In the first example, $4\cos^2(x)+7\sin x-7=0$, you can spot that $\cos^2(x)\equiv 1-\sin^2(x)$. Because the other function is sine, that looks promising!

   Similarly, in the second example, $2\cot^2(\theta)+3\csc(\theta)-7=0$, you can use $\cot^2(\theta)\equiv\csc^2(\theta)-1$.

2. **Replace the squared trig function.**

   The first example becomes

   $$4\left(1-\sin^2(x)\right)+7\sin(x)-7=0$$

   And the second becomes

   $$2\left(\csc^2(\theta)-1\right)+3\csc(\theta)-7=0$$

3. **Tidy up, being extremely careful with the brackets!**

   The first example is

   $$-4\sin^2(x)+7\sin(x)-3=0$$

   And you can flip the signs to make it

   $$4\sin^2(x)-7\sin(x)+3=0$$

   The second example is

   $$2\csc^2(\theta)+3\csc(\theta)-9=0$$

These equations are both now in a nicer form, which you can solve.

TIP

The examiners may throw in a curveball by offering something that's not in quite such an obvious form at first – for example, by putting in a $\tan(x)\cos(x)$ term that's equivalent to $\sin(x)$. Making things as simple as possible as quickly as possible is usually a good plan; if you see a mess, try to tidy it up!

Solving the nicer form

When you have everything in terms of one trig function, congratulations! That's a disguised quadratic. Here's how you tackle disguised quadratics. At this point, you have $4\sin^2(x)-7\sin(x)+3=0$ for the Core 2 example and $2\csc^2(\theta)+3\csc(\theta)-9=0$ for the Core 3 one.

1. **Create a substitute variable.**

   For the first example, pick $y = \sin(x)$, and for the second, use $u = \csc(\theta)$. (I've used different letters, but there's no significance in that – use whatever spare letters take your fancy.)

2. **Rewrite your equations using the new variable.**

   The first example is now

   $$4y^2 - 7y + 3 = 0$$

   And the second is

   $$2u^2 + 3u - 9 = 0$$

3. **Solve the equations.**

   These both factorise: the first is $(4y-3)(y-1)=0$, giving you $y = \frac{3}{4}$ or $y = 1$. The second is $(2u-3)(u+3)=0$, so $u = \frac{3}{2}$ or $u = -3$.

4. **Now solve for the original variables.**

   If $\sin(x) = \frac{3}{4}$, then $x \approx 0.848$ or $x \approx 2.294$ in the given domain. If $\sin(x) = 1$, $x = \frac{1}{2}\pi$ only. The answers to the first question are roughly 0.848, exactly $\frac{1}{2}\pi$ and roughly 2.294.

   If $\csc(\theta) = \frac{3}{2}$, then $\sin(\theta) = \frac{2}{3}$ and $\theta \approx 0.730$ or $\theta \approx 2.412$. If $\csc(\theta) = -3$, then $\sin(\theta) = -\frac{1}{3}$ and $\theta \approx 3.481$ or $\theta \approx 5.943$, giving you the four solutions to the second example.

WARNING!

Some of the values that come out of the quadratics may not give you a valid solution. Remember that sine and cosine give you values only between –1 and 1 inclusive, whereas secant and cosecant give you no values strictly between –1 and 1. If you see something like $\sin(\theta) = 3$, you can write 'no valid solutions' next to it to explain why you're not taking it any further.

More-involved disguises

It's possible (although unlikely) that you'll see a cubic or worse involving trigonometric functions. Don't panic. I repeat, do not panic: remain calm, and nobody will get hurt.

Remember two things: one, the examiners tend to walk you through unfamiliar questions, and two, they don't ask you anything you don't have the tools to deal with.

A typical question may ask you to solve $\tan^3(x) - \tan^2(x) - 3\tan(x) + 3 = 0$ for $0 \le x < 2\pi$. That would probably be considered a bit much straight off; most likely, the question would *start* by asking you to factorise $t^3 - t^2 - 3t + 3$, possibly even suggesting you show $t - 1$ is a factor.

Even without the hint, you can solve this problem using the following steps:

1. **Make a substitution for what's ugly.**

   Here, there are an awful lot of *tan*s taking up an awful lot of space, so try $t = \tan(x)$. You now have $t^3 - t^2 - 3t + 3 = 0$.

2. **Find a solution to your new polynomial.**

   If the examiners haven't given you a hint, you can expect the solutions to be easy enough to find. Trying a few values for t turns up $t = 1$ as a solution.

3. **Factorise and solve!**

   You have $(t - 1)(t^2 - 3) = 0$, which has solutions at $t = 1$, $t = \sqrt{3}$ and $t = -\sqrt{3}$.

4. **Solve for the original variable.**

   If $\tan(x) = 1$, then $x = \frac{1}{4}\pi$ or $x = \frac{5}{4}\pi$. If $\tan(x) = \sqrt{3}$, then $x = \frac{1}{3}\pi$ or $x = \frac{4}{3}\pi$. Lastly, if $\tan(x) = -\sqrt{3}$, then $x = \frac{2}{3}\pi$ or $x = \frac{5}{3}\pi$.

Changing the domain

It's not too tricky to make sure you have all the solutions when you have something as simple as $\sin(x)$. What if it's a monster like $\tan(3x + 40°) = 0.5$ for $0° \le x < 360°$? Then, dear reader, you have to get a bit creative. As usual, when you have something ugly, you replace it with something prettier. Here's how:

1. **Let θ (for example) be the ugly thing in the brackets.**

   Here, $\theta = 3x + 40$.

2. Change the domain so it's in terms of θ.

You simply do to the ends of the domain what you did to x – here, treble them and add 40°. If $0° \leq x < 360°$, then the lower end of the θ domain is $3 \times 0° + 40°$, or 40°; the upper end is $3 \times 360° + 40°$, or 1,120°. The domain for θ is $40° \leq \theta < 1,120°$.

3. Sketch your graph using the new limits.

In this case, you draw $y = \tan(\theta)$ from 40° to 1,120°.

4. Draw a horizontal line at the value you're trying to find. Then use your calculator and symmetry to find all the solutions for θ.

Your horizontal line hasn't moved; it's still $y = 0.5$, and your calculator should give you 26.6° as an answer (which is outside your new domain). You should also find $\theta \approx 206.6°$, 386.6°, 566.6°, 746.6°, 926.6° and 1,106.6°.

5. Now solve for the original variable, x.

If $\theta = 3x + 40°$, then the θ-values from Step 4 give you (approximate) x-values of 55.5°, 115.5°, 175.5°, 235.5°, 295.5° and 355.5°.

Adjusting the domain in this way 'magically' makes sure that the answers you find all land in the original domain when you convert them back to the original variable.

Chapter 13

Making Vectors as Simple as *i, j, k*

I have an opinion diametrically opposed to most A level students: I think there should be *much more* vector maths in the course. That's partly bias (a lot of my research was based on vectors) and partly pity: if you're lucky (or unlucky, depending on your point of view), you'll see three or four vector questions over the whole course, and they won't really be linked to anything else. I think that's a shame.

In its most abstract form, a vector is the combination of a distance with a direction. 'Ten metres to the north-east' is a perfectly good description of a vector, although A level, for the most part, doesn't delve into the abstract much. Instead, it deals with more concrete, coordinate-based vectors.

Variations on Vectors

You know what a *point* is: it's a fixed location in space. It might have coordinates: two or three of them, depending on how many dimensions you're working in. You can think of a *vector* as a map that takes you from one point to another. For example, to get from the point $(3, 4)$ to the point $(-7, 5)$, you'd need to move -10 in the x-direction and $+1$ in the y-direction. For reasons best known to historians, vectors are usually written in a column with the x-coordinate on top, like this: $\begin{pmatrix} -10 \\ 1 \end{pmatrix}$. See Figure 13-1a for a picture to make it clearer.

(a)

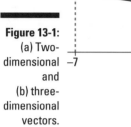

(b)

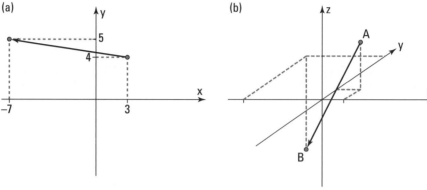

Figure 13-1:
(a) Two-
dimensional
and
(b) three-
dimensional
vectors.

Be careful: the vector from one point to the other isn't the same as the vector from the other point to the first; direction is important when it comes to vectors. The vector going back the other way would be $\begin{pmatrix} 10 \\ -1 \end{pmatrix}$ – the same thing, but with the signs reversed.

Opening your i's, j's and k's

You'll also see vectors written in '**i, j, k**' notation, like this: $-10\mathbf{i} + \mathbf{j}$. That means exactly the same thing, but it's easier to typeset. Algebraically, the **i**s, **j**s and **k**s (for the z-direction) work just the same as any other letters, but these special letters designate the axes, and you're usually more interested in the numbers on them for calculations.

Most people find it easier to use column vectors, so I'll be using those throughout this chapter. You're perfectly welcome to use **i, j, k** notation. You may notice that vectors are printed in **bold** text – obviously, it's hard to handwrite bold letters, so the convention when writing is to underline vectors rather than bold them.

Three-dimensional vectors work just the same way (see Figure 13-1b). To get from point A at $(1, 2, 3)$ to point B at $(-4, 5, -6)$, you need to go –5 in the x-direction, +3 in the y-direction and –9 in the z-direction, so your vector would be $\begin{pmatrix} -5 \\ 3 \\ -9 \end{pmatrix}$ or $-5\mathbf{i} + 3\mathbf{j} - 9\mathbf{k}$. You can show that the vector takes you from A to B by calling it $\overline{AB}$.

Vector language

Exam questions use language like 'relative to a fixed origin O', which is really just there for pedantic reasons – all it's saying is 'There's a point, we're calling it O for the origin, and it's going to be where we count from for the whole question.' The origin, conventionally, is at $(0, 0)$ or $(0, 0, 0)$, depending on how many dimensions you're in.

You may also see language like 'λ and μ are scalar parameters'. *Scalar* simply means 'a regular number' – in Core 4, a scalar is usually something you might multiply a vector by.

Lastly, a *component* of a vector is one of the numbers that make it up – the y-component of $\begin{pmatrix} 1 \\ 2 \\ 3 \end{pmatrix}$ is 2, for instance.

Recapping arithmetic

This seems a good point to remind you how to add and subtract vectors and how to multiply them by scalars: the short version is 'exactly as you'd expect'.

Adding a vector $\begin{pmatrix} a \\ b \\ c \end{pmatrix}$ to another, $\begin{pmatrix} d \\ e \\ f \end{pmatrix}$, gives you a third vector, $\begin{pmatrix} a+d \\ b+e \\ c+f \end{pmatrix}$. You simply add the components together.

Subtracting $\begin{pmatrix} a \\ b \\ c \end{pmatrix}$ from $\begin{pmatrix} d \\ e \\ f \end{pmatrix}$ gives you, exactly as you'd expect, $\begin{pmatrix} d-a \\ e-b \\ f-c \end{pmatrix}$; you subtract the components in the same order as you're subtracting the vector.

Finally, multiplying $\begin{pmatrix} a \\ b \\ c \end{pmatrix}$ by a scalar λ gives you the vector $\begin{pmatrix} \lambda a \\ \lambda b \\ \lambda c \end{pmatrix}$; each component in turn is multiplied by λ. It's worth noting in particular that

multiplying a vector by –1 reverses its direction.

Understanding magnitude

The *magnitude* of a vector simply means 'how large it is' – a vector is a line segment, so you can think of magnitude as the length of the vector. Your natural response, when you want to know how long something is in more than one dimension, is to reach for Pythagoras.

So if you're given a vector like $3\mathbf{i} + 12\mathbf{j} - 4\mathbf{k}$ (or $\begin{pmatrix} 3 \\ 12 \\ -4 \end{pmatrix}$) and you want to find its magnitude, here's what you do:

1. **Square each of the components.**

   You get 9, 144 and 16.

2. **Add them up.**

   That's 169.

3. **Take the positive square root of the answer.**

   The magnitude of the vector is 13.

You may be asked explicitly for the magnitude of a vector, either in words or using the modulus sign you know from Core 3 (for example, $|\mathbf{a}|$ and $\left|\overrightarrow{AB}\right|$ both mean 'the magnitude of the vector'); you may also be asked for the length of a vector or the distance between two points – those, too, are questions about the magnitude.

If you know how to find the magnitude of a vector, you nearly know how to find the distance between two points with known coordinates – let's say $(1,2,3)$ and $(8,7,-3)$:

1. **Find the vector that takes you from one point to the other.**

   The simplest way to do this is to find difference between each pair of x-, y-, and z-coordinates. You get $\begin{pmatrix} 7 \\ 5 \\ -6 \end{pmatrix}$.

2. **Find the magnitude of this vector.**

   Squaring the components gives you 49, 25 and 36, which sum to 110. The distance between the two points is $\sqrt{110}$.

Generally, you find the vector that takes you from point A (where $\overrightarrow{OA} = \mathbf{a}$) to point B (where $\overrightarrow{OB} = \mathbf{b}$) by working out $\mathbf{b} - \mathbf{a}$.

Lines in 3D: Writing Equations in Vector Form

You're familiar with the idea of a line in two dimensions (if not, Chapter 4 may be a good one to revisit). To define a 2D line, you need two things: a *reference point* (somewhere it goes through) and a *direction* (which, in 2D, is usually a gradient). For example, a line through the point $(-2, 3)$ with a gradient of 4 would have an equation of $(y - 3) = 4(x + 2)$.

The vector equation of a line follows the same idea, although it's written differently. This same 2D line could have the equation $\begin{pmatrix} x \\ y \end{pmatrix} = \begin{pmatrix} -2 \\ 3 \end{pmatrix} + \lambda \begin{pmatrix} 1 \\ 4 \end{pmatrix}$ – and it's made up of exactly the same parts!

The x and y refer to the coordinates, as you'd expect; the next bit is the reference point, $(-2, 3)$. The last bit is the slightly complicated bit: a gradient of 4 means that for each unit increase in the x direction, you get a four-unit increase in the y direction – which is exactly what the last column vector tells you. The λ simply means 'any multiple of this' – it's a parameter that works much like the x in a straight line equation.

A three-dimensional line works just the same way, only with an extra row. In this section, I guide you on how to find the equation of a line in 3D, find points on a line and work out whether (and where) two lines cross.

Writing the equation of a line through two points

To get the equation of a vector line, you need a point on the line, which tells you where the line is, and a direction vector, which tells you where it's going. The usual formula for a vector line passing through A and B is in the form $\mathbf{r} = \overrightarrow{OA} + \lambda \overrightarrow{AB}$, which needs a little explaining. There are four bits to it:

✔ **r** is the position vector of a general point on the line, which you can write as $\begin{pmatrix} x \\ y \\ z \end{pmatrix}$ if you prefer.

✔ $\overrightarrow{OA}$ is the position vector of point A – which makes sense; it's the vector that takes you from the origin O to point A.

✔ λ is a scalar parameter (see the sidebar 'Vector language'). It means 'whatever number you like', and I'll come back to it in a moment.

✔ $\overrightarrow{AB}$ is the vector from A to B.

Translating all this from maths into almost-normal-person-speak, it says, 'You start at O and then go to A. Then you go some distance (λ) in the direction from A to B. Whatever λ you pick, you end up on the line.' Figure 13-2 tries to convince you of this.

Figure 13-2:
Vector lines.

© John Wiley & Sons, Inc.

One way to see that this equation defines the line through A and B is to recognise it's a line (it keeps going in the same direction, so it's straight) and that it goes through A (when $\lambda = 0$) and through B (when $\lambda = 1$).

So to find the vector equation of a line through, for example, $(2, 3, 4)$ and $(-5, 2, 9)$, you do the following:

1. **Write down the general equation of a vector line.**

$$\mathbf{r} = \overrightarrow{OA} + \lambda \overrightarrow{AB}$$

2. **Work out $\overrightarrow{OA}$.**

The position vector of the first point is $\begin{pmatrix} 2 \\ 3 \\ 4 \end{pmatrix}$.

3. **Work out $\overrightarrow{AB}$.**

The vector between the given points is $\begin{pmatrix} -7 \\ -1 \\ 5 \end{pmatrix}$.

4. **Replace the bits that need replacing.**

$$\mathbf{r} = \begin{pmatrix} 2 \\ 3 \\ 4 \end{pmatrix} + \lambda \begin{pmatrix} -7 \\ -1 \\ 5 \end{pmatrix}$$

There's the equation of your line! Strictly speaking, it's *an* equation of the line – you could start from any point on the line, and your direction vector could be any multiple of $\begin{pmatrix} -7 \\ -1 \\ 5 \end{pmatrix}$ – but this is the one they're expecting to see.

(It's also acceptable to write the equation as $\mathbf{r} = 2\mathbf{i} + 3\mathbf{j} + 4\mathbf{k} + \lambda(-7\mathbf{i} - \mathbf{j} + 5\mathbf{k})$ or several variations on that theme.)

WARNING!

Using a different starting point or a different multiple of $\overrightarrow{AB}$ will give you a different equation of the same line – although you get the same collection of points whichever version you use, the same value of λ will generally give you different points in different equations.

Finding the equation of a line in a given direction

Finding a line parallel to a vector when given a point on the line is even easier than finding the equation of a line between two points; you can recycle the parallel vector as the direction of your line. Because you know a point on the line, it's just a case of substituting your two vectors into the equation of the line, $\mathbf{r} = \overrightarrow{OA} + \lambda\overrightarrow{AB}$.

In this case, $\overrightarrow{OA}$ tells you how to get to the point on the line, and $\overrightarrow{AB}$ is the parallel vector. The line through $(3, -1, 5)$ parallel to the vector $\begin{pmatrix} -1 \\ 3 \\ -2 \end{pmatrix}$ is

$\mathbf{r} = \begin{pmatrix} 3 \\ -1 \\ 5 \end{pmatrix} + \lambda \begin{pmatrix} -1 \\ 3 \\ -2 \end{pmatrix}$. It's really that simple!

Make sure you remember the λ – it's easy to overlook when you're gloating about how easy it is!

Showing that a point is on a given line

Showing that a point lies on a line is as simple as finding the value of the parameter λ (or whatever letter it is) that gives your point's coordinates as the output. For example, to show that the point $(3, 6, -2)$ lies on the line
$\mathbf{r} = \begin{pmatrix} -5 \\ 2 \\ 6 \end{pmatrix} + \lambda \begin{pmatrix} 2 \\ 1 \\ -2 \end{pmatrix}$, you first solve the x-direction equation: $3 = -5 + 2\lambda$, so $\lambda = 4$.

Then check the other two directions by putting your λ-value into the equation: in the y-direction, $2 + 4(1) = 6$, which is what you need; in the z-direction, $6 + 4(-2) = -2$, which is also the coordinate you need. That shows that the point is on the line!

You may be asked *whether* a point lies on a line. For example, does the point $(-7, 1, 7)$ lie on the same line $\mathbf{r}$? Using the same technique, you solve the x-direction equation to get $-7 = -5 + 2\lambda$, so $\lambda = -1$. Substituting this into the y-direction equation gives you $2 + 1(-1) = 1$, as needed; putting it into the z-direction equation gives you $6 + (-2)(-1) = 8$, which *isn't* the coordinate you were given – so this point does *not* lie on the given line.

If you're asked to show that the three points A, B and C at $(5, 1, 3)$, $(3, 2, 0)$, and $(-1, 4, -6)$ are co-linear (or collinear, depending on your taste), all you need to do is this:

1. **Find the vector from any one of the points to any other.**

    For example, you can say that the vector $\overrightarrow{AB} = \begin{pmatrix} -2 \\ 1 \\ -3 \end{pmatrix}$.

2. **Find the vector from any point to any other, but pick a different pair.**

    You might choose $\overrightarrow{BC} = \begin{pmatrix} -4 \\ 2 \\ -6 \end{pmatrix}$.

3. **If the result from Step 1 is a multiple of the result from Step 2, the three points are co-linear.**

    Here, $\overrightarrow{AB} = \frac{1}{2}\overrightarrow{BC}$, and the three points lie in a straight line.

This approach also gives you the ratio of distances between the points. Here, you have that $AB : BC = 1 : 2$.

Determining where (and whether) lines cross

For two lines to cross, they need to be in the same place at the same time, which means simultaneous equations. Having the equations in vector form is a little more complicated than regular simultaneous equations, as with vector form, you usually have three equations with two unknowns. That means you need to check your answers for the first two equations work for the third!

Whether two lines cross

To find out whether two lines cross, you set up three simultaneous equations: one each in the x-, y- and z-directions. If you can find values of λ and μ (or whatever your parameters are) that make all three equations work, then yes, they cross! If you can't? Then they don't cross. Here's an example:

With respect to a fixed origin O, the lines l_1 and l_2 are given by the equations:

$l_1 : \quad \mathbf{r} = (-7\mathbf{i} + \mathbf{j} + 9\mathbf{k}) + \lambda(2\mathbf{i} + \mathbf{j} - \mathbf{k})$
$l_2 : \quad \mathbf{r} = (6\mathbf{i} + 22\mathbf{k}) + \mu(3\mathbf{i} - \mathbf{j} + 5\mathbf{k})$

Show that l_1 and l_2 meet.

1. **Optionally, rewrite the equations in column vector form.**

   It's easier to see the equations you need if you write them as
   $$\mathbf{r} = \begin{pmatrix} -7 \\ 1 \\ 9 \end{pmatrix} + \lambda \begin{pmatrix} 2 \\ 1 \\ -1 \end{pmatrix} \text{ and } \mathbf{r} = \begin{pmatrix} 6 \\ 0 \\ 22 \end{pmatrix} + \mu \begin{pmatrix} 3 \\ -1 \\ 5 \end{pmatrix}.$$

2. **Equate the *x*-, *y*- and *z*-components of each line.**

   $-7 + 2\lambda = 6 + 3\mu$ in the *x*-direction, $1 + \lambda = -\mu$ in the *y*-direction, and $9 - \lambda = 22 + 5\mu$ in the *z*-direction.

3. **Pick the pair that looks easiest to solve and solve them for μ and λ.**

   The *x* and *y* equations look relatively nice. I'd substitute $\lambda = -1 - \mu$ into the first equation to get $-7 + 2(-1 - \mu) = 6 + 3\mu$; $\mu = -3$ with a bit of rearranging. Then $\lambda = -1 + 3 = 2$.

4. **Check whether these values work in the third equation.**

   Does $9 - \lambda = 22 + 5\mu$? The left-hand side is 7, the right-hand side is also 7, so they match up, which means the lines *do* intersect.

Had the final equation not matched up in this question, you'd know that you'd made a mistake (because you were asked to show that the lines cross). However, if you're asked to find out *whether* two (non-parallel) lines cross and the last equation doesn't match up, then the lines don't intersect – the two lines are *skew*.

You can think of the two lines as bits of string stretched across the room. If you look down from above, they may appear to cross, but you need to check the last coordinate to see whether they're in the same place vertically. That's what you're doing with this check.

Where two lines cross

Finding *where* two lines cross is almost identical to finding out whether they cross, except with one extra step: you need to find the coordinates given by putting λ and μ into the original line equations. (It's good practice to check they give the same answer.)

For the example from the preceding section, putting $\lambda = 2$ into the original equation for l_1 gives you $\mathbf{r} = \begin{pmatrix} -3 \\ 3 \\ 7 \end{pmatrix}$; putting $\mu = -3$ into l_2 gives you the same result. The lines cross at $(-3, 3, 7)$.

Picking constants so that lines cross

Related to the crossing-lines questions is the kind of question that defines two lines using an unknown constant. The challenge is to find the constant that makes the lines intersect. For example, two lines may be defined as follows:

$$\mathbf{r_1} = \begin{pmatrix} 2 \\ -4 \\ -1 \end{pmatrix} + \lambda \begin{pmatrix} 5 \\ 4 \\ -2 \end{pmatrix}$$

$$\mathbf{r_2} = \begin{pmatrix} a \\ 6 \\ 3 \end{pmatrix} + \mu \begin{pmatrix} 5 \\ -14 \\ -2 \end{pmatrix}$$

If you're told they intersect, what does the constant a need to be? You have to solve three simultaneous equations – although, luckily, two have only λ and μ in them, so you can solve them and use the answers to find a in the remaining equation.

Here, you solve the y and z equations, $-4 + 4\lambda = 6 - 14\mu$ and $-1 - 2\lambda = 3 - 2\mu$, to get $\mu = 1$ and $\lambda = -1$. Putting those values into the x equation gives you $2 + 5(-1) = a + 5(1)$, so $a = -8$.

Whenever you're asked to find the value of a constant, treat it like a number and use algebra to solve it in the end.

Dot Products: Multiplying Vectors

There are two (sensible) ways of multiplying vectors together. One, the *vector product* or *cross product*, gives you a vector as a result. You probably won't care about the vector product unless and until you do Further Maths, so I'll say no more about it. Meanwhile, the *scalar product* or *dot product* gives you a number (a scalar) as a result, and it's much more useful for Core 4.

There are two ways of defining the dot product, and you'll need both (usually at the same time). If you have two vectors, $\mathbf{a} = \begin{pmatrix} a_x \\ a_y \\ a_z \end{pmatrix}$ and $\mathbf{b} = \begin{pmatrix} b_x \\ b_y \\ b_z \end{pmatrix}$,

then $\mathbf{a} \cdot \mathbf{b} = a_x b_x + a_y b_y + a_z b_z$. That is, the dot product is the product of the x-components of each vector, the product of the y-components and the product of the z-components, all added up. The other definition is less natural but just as useful: $\mathbf{a} \cdot \mathbf{b} = |\mathbf{a}||\mathbf{b}|\cos\theta$, where θ is the angle between the vectors and a pair of vertical bars represents the magnitude of the vector inside them.

Showing vectors are perpendicular

A common use of the dot product is to show that two vectors or lines are perpendicular. (Two lines are perpendicular if their direction vectors are perpendicular.) The dot product of perpendicular vectors is always 0. Why? It's because of the second definition of the dot product $-\mathbf{a}\cdot\mathbf{b}=|\mathbf{a}||\mathbf{b}|\cos\theta-$ and because $\cos(90°)=0$. (For reasons best known to the examiners, vector questions almost always use degrees instead of radians.)

For example, suppose you have two lines defined as $\mathbf{r_1}=\begin{pmatrix}1\\2\\3\end{pmatrix}+\lambda\begin{pmatrix}3\\-8\\5\end{pmatrix}$ and $\mathbf{r_2}=\begin{pmatrix}0\\-5\\13\end{pmatrix}+\mu\begin{pmatrix}-3\\2\\5\end{pmatrix}$, and you need to show that they're perpendicular. Here's the recipe:

1. **If you're dealing with lines, write down the direction vectors.**

   Here, they're $\begin{pmatrix}3\\-8\\5\end{pmatrix}$ and $\begin{pmatrix}-3\\2\\5\end{pmatrix}$.

2. **Work out the dot product of the vectors.**

   You get $(3)(-3)+(-8)(2)+(5)(5)=-9-16+25=0$.

3. **If the dot product is 0, the vectors are perpendicular.**

   It is. They are.

Finding constants that give you perpendicular vectors uses the same trick: if you have vectors such as $\begin{pmatrix}1\\2\\3\end{pmatrix}$ and $\begin{pmatrix}-1\\2\\k\end{pmatrix}$, and you need to find the value of k that makes them perpendicular, simply find the dot product, which is $(1)(-1)+(2)(2)+3(k)=3+3k$, and set the result equal to 0. If $3+3k=0$, then $k=-1$.

In three (or more) dimensions, lines can be perpendicular without intersecting. For instance, any vertical line is perpendicular to any horizontal line, even if they don't have any points in common. Your chances of seeing this in an exam are very slim, and your chances of having to *do* anything related to it are slimmer still.

Finding angles between vectors

Much more common than perpendicular questions are angular questions: given two vectors (or two lines), find the angle between them. That's a bit nastier, but it boils down to using the two definitions of the dot product and solving for θ.

Imagine you need to find the acute angle, in degrees, between the vectors $\mathbf{u} = \begin{pmatrix} -2 \\ 1 \\ 2 \end{pmatrix}$ and $\mathbf{v} = \begin{pmatrix} 3 \\ -3 \\ 0 \end{pmatrix}$. Here's how you do it:

1. **If you're dealing with lines, get the direction vectors.**

    Here, you already have the vectors.

2. **Work out the dot product of the vectors.**

    $$\mathbf{u} \cdot \mathbf{v} = (-2)(3) + (1)(-3) + (2)(0) = -9$$

3. **Work out the modulus of each vector.**

    $$|\mathbf{u}| = \sqrt{4+1+4} = 3$$
    $$|\mathbf{v}| = \sqrt{9+9+0} = 3\sqrt{2}$$

4. **Now use the other dot-product definition.**

    $$\mathbf{u} \cdot \mathbf{v} = |\mathbf{u}||\mathbf{v}|\cos(\theta)$$

    You have $-9 = 9\sqrt{2}\cos(\theta)$, so $\cos(\theta) = -\dfrac{1}{\sqrt{2}}$, and $\theta = 135°$.

5. **Make sure you give the angle they ask for!**

    If two lines cross at 135°, they also cross at 45°, which is the acute angle they're after.

Answering Evil Vector Questions

There's a cartoon somewhere about a geography exam question that asks '1a) Approximately how many people live in the UK? (2 marks). 1b) Name them. (6 marks).' The Core 4 vector questions often seem to be set up in a similar way: the first half is simple and fairly predictable, and then they unleash hordes of hungry vampires for the second half.

Luckily, you have your garlic and holy water; you have prepared yourself for the onslaught and will be able to defeat the vector vampires even without a cross product.

Sorry? No, I'm not sorry. Why do you ask?

Points on perpendicular lines and shortest distances

About 30 per cent of the Core 4 papers I've seen ask you a horrible perpendicular point question. They give you a line and point P, and they want you to find point Q on the line such that $\overrightarrow{PQ}$ is perpendicular to the line.

For example, take P as $(4, 16, -3)$ and the line as $\mathbf{r} = \begin{pmatrix} 6 \\ 1 \\ 3 \end{pmatrix} + \lambda \begin{pmatrix} 1 \\ 4 \\ -2 \end{pmatrix}$. Here's the drill:

1. **Work out $\overrightarrow{PQ}$.**

   Whatever point Q is, it lies on the line. That means its coordinates are $(6 + \lambda, 1 + 4\lambda, 3 - 2\lambda)$, using the equation of the line. Vector $\overrightarrow{PQ}$ is the map from P to Q, which you can find by subtracting the coordinates of P from the coordinates of Q, leaving you $\overrightarrow{PQ} = \begin{pmatrix} 2 + \lambda \\ -15 + 4\lambda \\ 6 - 2\lambda \end{pmatrix}$.

2. **Find the dot product of $\overrightarrow{PQ}$ with the direction vector of the line.**

   The direction vector is $\begin{pmatrix} 1 \\ 4 \\ -2 \end{pmatrix}$, so the dot product is

   $$(2 + \lambda)(1) + (-15 + 4\lambda)(4) + (6 - 2\lambda)(-2)$$
   $$= 2 + \lambda - 60 + 16\lambda - 12 + 4\lambda$$
   $$= 21\lambda - 70$$

3. **Set the dot product equal to 0 and solve for λ.**

   Because $\overrightarrow{PQ}$ is perpendicular to the line, the dot product must be 0: $21\lambda - 70 = 0$, so $\lambda = \dfrac{10}{3}$.

4. Use the equation of the line to find Q.

$$\begin{pmatrix} 6+\dfrac{10}{3} \\[1ex] 1+\dfrac{40}{3} \\[1ex] 3-\dfrac{20}{3} \end{pmatrix} = \begin{pmatrix} \dfrac{28}{3} \\[1ex] \dfrac{43}{3} \\[1ex] -\dfrac{11}{3} \end{pmatrix}$$

Therefore, point Q is at $\left(\dfrac{28}{3}, \dfrac{43}{3}, -\dfrac{11}{3} \right)$.

If you're asked for the shortest distance from the point to the line, it's just the magnitude of this perpendicular vector.

Points at a given distance

Once in a while, you're asked to find points on a line at a given distance from another point. For example, given point A at $(6, 4, 5)$, find points P and Q on the line $\mathbf{r} = \begin{pmatrix} 7 \\ 6 \\ -11 \end{pmatrix} + \lambda \begin{pmatrix} 5 \\ 1 \\ -8 \end{pmatrix}$ such that $\left| \overrightarrow{AP} \right| = \left| \overrightarrow{AQ} \right| = 9$.

The plan is somewhat similar to the example in the preceding section: you can figure out the coordinates of P or Q in terms of λ and use what you know to calculate λ. Here's how:

1. Find an expression for, say, $\overrightarrow{AP}$.

Point P is on the line, so its coordinates are $(7+5\lambda, 6+\lambda, -11-8\lambda)$ for some value of λ. Vector $\overrightarrow{AP}$ is the map from A to P, which you get by subtracting the coordinates of A from those of P, giving you $\begin{pmatrix} 1+5\lambda \\ 2+\lambda \\ -16-8\lambda \end{pmatrix}$.

2. You're trying to find λ such that $\left| \overrightarrow{AP} \right| = 9$, so work out the magnitude of $\overrightarrow{AP}$.

The magnitude is

$$\sqrt{(1+5\lambda)^2 + (2+\lambda)^2 + (-16-8\lambda)^2}$$
$$= \sqrt{1+10\lambda+25\lambda^2+4+4\lambda+\lambda^2+256+256\lambda+64\lambda^2}$$
$$= \sqrt{90\lambda^2+270\lambda+261}$$

3. Find the values of λ that make the magnitude correct.

Squaring both sides of $\sqrt{90\lambda^2 + 270\lambda + 261} = 9$ gives you $90\lambda^2 + 270\lambda + 261 = 81$, or $90\lambda^2 + 270\lambda + 180 = 0$. That simplifies to $90(\lambda+1)(\lambda+2) = 0$, so $\lambda = -1$ or $\lambda = -2$.

4. Put these values of λ back into the line equation to find the points.

You get $(2, 5, -3)$ and $(-3, 4, 5)$.

Even though you arbitrarily picked point P to start with, the workings came up with two possible values of λ, corresponding to the two points on the line that work. These are P and Q, in either order.

Reflections in a line

Questions about reflections in a line are unusual in Core 4, but they have cropped up. If two lines intersect at right angles at point X, say, at $(-3, 1, 5)$, and you have point P, $(0, 4, 1)$, on one line, you can reflect it in the other line. The reflection gives you a point on the first line as far from X as P originally was but on the other side of X. You work the problem as follows:

1. Find the vector from your point to the intersection, $\overrightarrow{PX}$. Note that the direction is important!

Here, $\overrightarrow{PX} = \begin{pmatrix} -3 \\ -3 \\ 4 \end{pmatrix}$.

2. Add this displacement on to X to find the coordinates of the new point.

This gives you $(-6, -2, 9)$. The resulting point is the reflection of P in the line.

Areas of shapes

When working with vector shapes, you need to remember that the area of a triangle is $\frac{1}{2}bh$ or $\frac{1}{2}ab\sin(C)$, depending on what information you have, and that a parallelogram... well, it's just two triangles stuck together, so it's bh or $ab\sin(C)$, where h is the perpendicular height. The trick is to find the information. Read on!

Triangles

You'll see two types of triangle area questions: one where you know you have perpendicular vectors, and one where you don't.

If you know that two of the vectors making up the triangle are perpendicular –
for example, because you've worked it out earlier in the question – your
mission is clear: find the lengths of the two perpendicular vectors. These
become your base and your height, which you multiply together and halve to
find the area. Ta-da!

If you don't know there's a right angle, then your work is a bit more
involved. Say your points are at $A(1, 2, 3)$, $B(2, -1, 2)$, and $C(0, 3, 0)$. Here's
what you do:

1. **Pick two of the vectors that make up the triangle. It's important
   to make sure they start at the same point – again, the direction is
   important.**

   Let's go for $\overrightarrow{CB}$, which is $\begin{pmatrix} 2 \\ -4 \\ 2 \end{pmatrix}$, and $\overrightarrow{CA}$, which is $\begin{pmatrix} 1 \\ -1 \\ 3 \end{pmatrix}$.

2. **Find their magnitudes.**

   These will be $|\mathbf{a}|$ and $|\mathbf{b}|$ in your dot-product equation, $\mathbf{a} \cdot \mathbf{b} = |\mathbf{a}||\mathbf{b}|\cos\theta$,

   and a and b in your triangle-area equation, Area $= \frac{1}{2}ab\sin\theta$. You get
   $|\mathbf{a}| = \sqrt{24}$ and $|\mathbf{b}| = \sqrt{11}$.

3. **Find the angle between the vectors (use the recipe from the earlier
   section 'Finding angles between vectors').**

   This is angle θ:

   $$\overrightarrow{CA} \cdot \overrightarrow{CB} = 12$$
   $$\sqrt{24}\sqrt{11}\cos(\theta) = 12$$
   $$\cos(\theta) = \frac{12}{\sqrt{24}\sqrt{11}}$$

   That gives you $\theta = 42.39°$, to two decimal places.

4. **Stick a, b and C into $\frac{1}{2}ab\sin(C)$, and presto – one triangle area.**

   The area is $\frac{1}{2}\sqrt{24}\sqrt{11}\sin(42.39°) = 5.48$ units^2.

It's unusual to see a problem like this – and if you do, you'll likely be walked
through finding the angle and magnitudes. All the same, it's handy to have
the know-how in case either it *does* come up or if you get a question where
you don't spot the right angle.

Parallelograms and rectangles

You may be asked to do a couple of things with a parallelogram (of which a rectangle is a special case; everything that's true of a parallelogram is also true of a rectangle).

You may need to find the coordinates of a missing corner, given the three others. Here, you use the fact that the vectors on opposite sides of the parallelogram are equal (they're parallel – that's where the name *parallelogram* comes from – and they're the same length, so they must be the same vector). Suppose that the parallelogram is *ABCD* and that *C* is the unknown point. A sketch (see Figure 13-3) shows that $\overrightarrow{AB} = \overrightarrow{DC}$. You can work out $\overrightarrow{AB}$ from the given coordinates; you use the same map to get from *D* to *C* as you use to get from *A* to *B*, so moving from *D*'s coordinates gives you the coordinates of *C*.

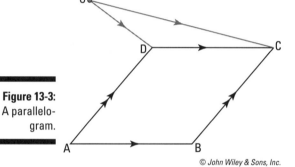

Figure 13-3:
A parallelo-
gram.

© John Wiley & Sons, Inc.

Finding the area of a parallelogram given three points is straightforward, too. Here's the recipe, with spooky echoes of the triangle area recipe from the preceding section:

1. **Find two adjacent vectors that make up the parallelogram.**

2. **Find their magnitudes.**

   These will be $|\mathbf{a}|$ and $|\mathbf{b}|$ in the dot-product equation, $\mathbf{a} \cdot \mathbf{b} = |\mathbf{a}||\mathbf{b}|\cos\theta$, and *a* and *b* in the parallelogram area equation, Area $= ab\sin\theta$.

3. **Find the angle between the vectors using the dot-product equation.**

   This will be θ.

4. **The area of the parallelogram is $ab\sin(\theta)$.**

Miscellaneous evildoing

There's no end to the evil vector questions the examiners can ask, in honesty. I've seen ones where they give you a point, *P,* and a line and ask you to find the points *Q* and *R* on the line such that angles *PQR* and *PRQ* are both 45°. It wasn't *as* bad as it might have been, because you could use isosceles-triangle facts to turn the question into a points-at-a-distance question and solve it without getting into messy dot products!

Similarly, a question may note that a circle centred on a given point *C* cuts a line at point *A* and ask where else the line and circle intersect. This is a cunningly disguised point-at-a-distance question: you just need to find the distance from *A* to *C* and to find both points on the line at that distance from *C*. That will give you two answers, one of which will be *A*; you're interested in the other.

Back to Normal: Picking Out Planes

A quick heads-up: if you're studying for a board that *isn't* MEI, then this section is almost certainly of no interest to you, at least as far as A level goes; sadly, you'll be missing out on one of the nicest bits of vector maths: the different kinds of equations of a plane.

A *plane* isn't just what gets you to Barcelona; it's a mathematical idea, too: you can think of it as a flat surface extending as far as you like in any direction – for example, a sheet of cardboard is part of a plane. Your ceiling is (practically speaking) a plane.

In this section, I show you how to work out the equation of a plane, how to find where a line and a plane intersect, and how to find the angle between planes.

Equation of a plane

Like the equation of a line, the equation of a plane gives some relation that the coordinates of its points have to satisfy. For example, $z = 0$ defines a plane: in particular, it's the plane consisting of all points with a z-coordinate of 0. You'd find $(0, 0, 0)$ in this plane as well as $(1, 2, 0), (-7, 3.5, 0)$ and $(\pi, -94, 0)$; the point $(0, 0, 1)$ most definitely *isn't* in this plane, because its z-coordinate isn't 0.

A plane can be defined by any three points that lie in it (as long as they're not in a straight line), but in terms of finding a plane's equation, it's less involved to find just two things: the normal vector and another number that you can think of as a sort of 'intercept', like where a straight line crosses the y-axis

(although it's a bit more complicated than that – the details are not important to you here). The *normal* vector is perpendicular to the plane – for example, if you think of the floor as a plane, the normal vector points straight up.
If you think of this page of the book as a plane (please hold it flat, thank you),
any vector pointing out of it, probably straight at you, is a normal vector. I
show this in Figure 13-4.

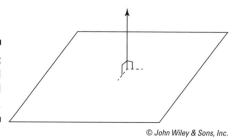

Figure 13-4:
A plane and
its normal
vector.

In this section, I show you how to figure out the normal vector and how to
write the equation of a plane in two different ways: the Cartesian way (which
involves *x, y* and/or *z*) and the vector way (which involves vectors, surprisingly).

Finding the normal

You won't be asked to find the normal vector directly (that's a Further Maths
topic), but you may well be asked to show that a given vector is normal
(perpendicular) to a plane.

For example, you may be given the points $A\,(1, 3, 2)$, $B\,(0, 5, 1)$ and $C\,(13, 0, 0)$
and be asked to show that the vector $\begin{pmatrix} 1 \\ 2 \\ 3 \end{pmatrix}$ is normal to the plane containing
them. That's not especially tricky – you just need to show that the given
vector is perpendicular to two vectors using the three given points. (There
are Good Reasons for this, but they're beyond the scope of this book.)

1. **Work out two nonparallel vectors using the three given points – for
   example, $\overrightarrow{AB}$ and $\overrightarrow{AC}$.**

   You have $\overrightarrow{AB} = \begin{pmatrix} -1 \\ 2 \\ -1 \end{pmatrix}$ and $\overrightarrow{AC} = \begin{pmatrix} 12 \\ -3 \\ -2 \end{pmatrix}$.

2. **Find the dot product of each of these vectors with the given vector.**

   The first gives you $-1 + 4 - 3 = 0$, and the second gives you $12 - 6 - 6 = 0$.

3. **Note that because you've shown that two nonparallel vectors that lie in the plane are perpendicular to the given vector, the vector is normal to the plane.**

The Cartesian equation of a plane

The *Cartesian equation* of a plane looks something like $ax + by + cz = d$, where a, b, c and d are constants. In particular, $\begin{pmatrix} a \\ b \\ c \end{pmatrix}$ is the normal vector to the plane, and d is an 'intercept' you need to work out. How do you do that? Simple: you use a point you know, just like you'd find the missing constant when integrating (see Chapter 16). For example, if point $A(2, 4, 7)$ is in a plane whose normal vector is $\begin{pmatrix} 1 \\ -5 \\ 2 \end{pmatrix}$, here's what you'd do:

1. **Write down each component of the normal vector followed by the coordinate axis it corresponds to, followed by '$= d$' – which is a constant you need to find.**

   $x - 5y + 2z = d$

2. **Put the point that you know is on the plane into the equation as x, y and z.**

   This gives you d:

   $d = 2 - 5 \times 4 + 2 \times 7 = -4$

3. **Write down the final equation.**

   $x - 5y + 2z = -4$

To check whether a point lies in this plane, you simply put its coordinates into the equation. If the equation holds true, then the point lies in the plane; if it doesn't, then the point doesn't lie in the plane.

The vector equation of a plane

The vector equation of a plane looks like this: $\mathbf{r} \cdot \mathbf{n} = d$, where $\mathbf{r}$ is the general vector $\begin{pmatrix} x \\ y \\ z \end{pmatrix}$, like for vector lines, $\mathbf{n}$ is the specific normal vector, and d is the 'intercept', worked out just like in the preceding section. Taking the same example (where point $A\,(2,4,7)$ is in a plane whose normal vector is $\begin{pmatrix} 1 \\ -5 \\ 2 \end{pmatrix}$), here's how you write the vector equation of the plane:

1. **Write down the template, replacing n with the normal vector.**

$$\mathbf{r} \cdot \begin{pmatrix} 1 \\ -5 \\ 2 \end{pmatrix} = d$$

2. **Work out _d_ by replacing r with the known point.**

$$d = \begin{pmatrix} 2 \\ 4 \\ 7 \end{pmatrix} \cdot \begin{pmatrix} 1 \\ -5 \\ 2 \end{pmatrix} = -4$$

Gosh, there's a coincidence.

3. **Write down the final equation of the plane.**

$$\mathbf{r} \cdot \begin{pmatrix} 1 \\ -5 \\ 2 \end{pmatrix} = -4$$

Of course, if you write **r** as $\begin{pmatrix} x \\ y \\ z \end{pmatrix}$ and work out the dot product explicitly, you get the Cartesian plane equation.

Plane intersections with lines

The natural thing to do when you have a plane is to figure out where it crosses a line and at what angle. There's a nice, logical way to do that!

Most vector problems boil down to simultaneous equations, Pythagoras or the dot product. The main difficulty is deciding between the options!

Coordinates of the intersection

Right-o: you've got the equation of a plane, something like $2x + 3y + 5z = 10$, and the vector equation of a line, something like $\mathbf{r} = \begin{pmatrix} 3 \\ -5 \\ -3 \end{pmatrix} + \lambda \begin{pmatrix} -1 \\ 3 \\ 2 \end{pmatrix}$. Where do they intersect? A good approach is to find the value of λ that makes the point on the line satisfy the plane equation. Clear? Well, it will be.

1. **Find expressions for x, y and z from the line equation.**

   You have $x = 3 - \lambda$, $y = -5 + 3\lambda$ and $z = -3 + 2\lambda$.

2. **Substitute these expressions for x, y and z in the plane equation.**

   You have $2(3 - \lambda) + 3(-5 + 3\lambda) + 5(-3 + 2\lambda) = 10$.

3. **Simplify and solve for λ.**

$$(6 - 2\lambda) + (-15 + 9\lambda) + (-15 + 10\lambda) = 10$$
$$-24 + 17\lambda = 10$$
$$\lambda = 2$$

4. **Now put this value into the line equation to find the point of intersection.**

   When $\lambda = 2$, $\mathbf{r} = \begin{pmatrix} 1 \\ 1 \\ 1 \end{pmatrix}$. Therefore, the line intersects the plane at $(1, 1, 1)$.

Note that $\mathbf{r}$ is the same thing as $\begin{pmatrix} x \\ y \\ z \end{pmatrix}$ when you're dealing with three-dimensional vectors.

Angles between intersecting lines and planes

Finding the angle between a line and a plane requires a little bit of care, but it's not as difficult as you may think. Here's what you'd do to find the angle between $2x + 3y + 5z = 10$ and $\mathbf{r} = \begin{pmatrix} 3 \\ -5 \\ -3 \end{pmatrix} + \lambda \begin{pmatrix} -1 \\ 3 \\ 2 \end{pmatrix}$ (the plane and line from the preceding section):

1. **Find the normal vector to the plane and the direction vector of the line.**

   Here, they're $\begin{pmatrix} 2 \\ 3 \\ 5 \end{pmatrix}$ and $\begin{pmatrix} -1 \\ 3 \\ 2 \end{pmatrix}$.

2. **Find the angle between these vectors (see the earlier section 'Finding angles between vectors').**

   You need to use the dot product, noting that $\mathbf{a} \cdot \mathbf{b} = |\mathbf{a}||\mathbf{b}|\cos\theta$.

   The dot product is $(2)(-1) + (3)(3) + (5)(2) = 17$; the magnitudes are $\sqrt{4 + 9 + 25} = \sqrt{38}$ and $\sqrt{1 + 9 + 4} = \sqrt{14}$. That means the angle is $\cos^{-1}\left(\dfrac{17}{\sqrt{38 \times 14}}\right) \approx 42.5°$.

3. **Step 2 gives you the angle between the normal and the line, so the angle between the plane and the line is that answer subtracted from a right angle, as you can see in Figure 13-5.**

The final answer here is $90° - 42.5° = 47.5°$.

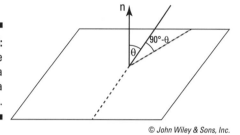

Figure 13-5:
The angle between a line and a plane.

© John Wiley & Sons, Inc.

Angles between one plane and another

If you need to find the angle between two planes, the recipe is much simpler than you'd expect: you simply find the angle between the normal vectors (see Figure 13-6).

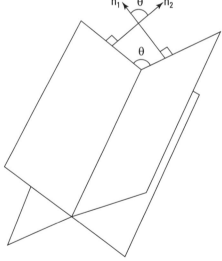

Figure 13-6:
The angle between two planes.

© John Wiley & Sons, Inc.

For example, to find the angle between $x + 2y + 3z = 4$ and $3x - 2y + z = 9$, you find the angle between $\begin{pmatrix} 1 \\ 2 \\ 3 \end{pmatrix}$ and $\begin{pmatrix} 3 \\ -2 \\ 1 \end{pmatrix}$ using the dot-product routine $\left(\mathbf{a} \cdot \mathbf{b} = |\mathbf{a}||\mathbf{b}|\cos\theta \right)$:

$$3 - 4 + 3 = \sqrt{1 + 4 + 9} \times \sqrt{9 + 4 + 1} \cos(\theta)$$

So $2 = 14\cos(\theta)$ and $\cos(\theta) = \frac{1}{7}$, making $\theta \approx 81.8°$.

Part IV
Calculus

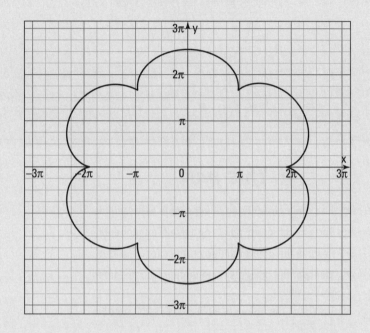

In this part . . .

- ✔ Find slopes by differentiating simple expressions.
- ✔ Sort out stationary points and go off on tangents.
- ✔ Reverse the differentiation process to integrate.
- ✔ Discover the calculus rules for more-complicated expressions.
- ✔ Turn back the forces of evil by answering horrible questions.

Chapter 14

Climbing Slippery Slopes

· ·

In This Chapter

▶ Understanding differentiation and its notation

▶ Applying the power rule

▶ Making use of the linear properties

▶ Differentiating trig, log and exponential functions

▶ Finding specific gradients

· ·

*T*ogether, *differentiation* and *integration* make up the subject of *calculus*, one of the great triumphs of seventeenth-century maths (and, according to some people, of all human history. I think that's a stretch). It's certainly true that the discovery of calculus dramatically opened up what could be done with maths, and we'd be in a much less interesting world without it.

You'll have to trust me on that, I'm afraid: the vast bulk of the calculus you do in A level is about finding the gradient of a slope and the area under a curve – although you do get to apply the techniques to 'real-world' problems, which you can find in Chapter 15.

In this chapter, I show you the basic techniques for differentiating.

Taking Slope to the Limit: What Differentiation Is

You've come across the idea of a gradient before at GCSE: it's the *m* bit in $y = mx + c$, and it's a measure of how steep a straight line is. If the gradient is a big positive number, the graph is quite steep and is going up as you go right; a smaller positive number means a shallower slope. A gradient of 0 is dead flat, whereas negative numbers mean downhill slopes: they go down as you go to the right.

REMEMBER

The gradient of the line segment between two points is the vertical difference between the points divided by the horizontal difference. You may remember it as 'the change in y over the change in x' or as 'rise over run' or even as $\frac{y_2 - y_1}{x_2 - x_1}$, if you like convoluted formulas. (Personally, I'm a fan of $\frac{\Delta y}{\Delta x}$, but for most people, that asks more questions than it answers.)

What differentiation does is extend that idea to the gradient of a *curve* – or, more precisely, the gradient of the tangent to a curve at any given point. If you think about a quadratic curve like $y = x^2$, you can see that the gradient changes as you go along it – for negative values of x, the curve is going downhill; at 0, the graph goes flat, and as x becomes positive, so does the gradient. But how can you put a number on a gradient that's continually changing?

What you're about to see isn't the simplest way to differentiate (that comes later in the chapter): this is an explanation of how the idea developed, and on some boards, you may be tested on it.

There's an apparent problem with finding a tangent line to a curve: you need two points to make a line, and you know only one. Luckily, there's a clever way around it, which is to take the point you're interested in and another point on the curve nearby. You then try the same thing for an even closer point and see how the gradient behaves.

For example, to get an approximation of the gradient of $y = x^2$ at a given value of x – let's say 3, so the point you're looking at is $(3, 9)$ – you can find the gradient of a line that goes through that point and another point on the curve nearby. Try $x = 3.1$, and you find the point $(3.1, 9.61)$; the line through $(3, 9)$ and this has a gradient of $\frac{0.61}{0.1} = 6.1$. A bit closer together, try $(3, 9)$ and $(3.001, 9.006001)$, which gives you a gradient of 6.001. You can generalise this further with a bit of algebra: if you make the second point $(3 + h, \ 9 + 6h + h^2)$, the gradient works out to be $\frac{6h + h^2}{h} = 6 + h$. (This fits precisely with the answers you've seen so far, by the way.) As the second point moves closer and closer to $x = 3$, h gets closer and closer to 0, and the gradient gets closer and closer to 6: it would be perverse not to say that the gradient of the curve at $x = 3$ was 6. Mathematicians say that 'in the limit as h goes to 0, the gradient goes to 6'.

This technique works on any reasonably well-behaved curve (certainly any curve you're likely to see at A level): to find the gradient at a given point $\left(x_0, \ f(x_0)\right)$, consider the gradient of the line to $\left(x_0 + h, \ f(x_0 + h)\right)$ and work out what happens as h becomes very small.

Luckily, mathematicians have saved you the trouble of doing this every time: for the functions you're going to come across, you can use simplifying rules and techniques.

In all of the above, I've taken h to be a positive number, but it works just as well if h is negative. In fact, getting the same answer with positive and negative h is part of what I mean by 'a reasonably well-behaved curve'; if this is the case at a particular value of x, a mathematician would say the curve was *differentiable at x*.

Designating Derivatives: A Note on Notation

There are several different ways of denoting a *derivative*, which is what you call the result of differentiating. (A *differential* is something slightly different, and you don't care about it; you just need to know that differentiating something gives you a derivative.)

First, you can use the 'prime' notation, due to Isaac Newton: you can denote the derivative of a variable by putting a straight little quote mark after it – the derivative of y with respect to x is y'. Similarly, if you differentiate a function, you can write $f'(x)$ to show the derivative. A second derivative (which you get by differentiating the derivative) gets another prime symbol, so it may look like y'' or $f''(x)$.

Unfortunately, this notation quickly gets hard to read, and it's easy to mistake a fleck of dirt on your paper or screen for a prime symbol (or vice versa), so it's not my preferred way of writing things down. (It's also limited by the number of variables it makes sense to use; although you can use a dot to denote derivatives with respect to time, you run out of symbols in no time and end up saying things like 'x double dot dot x dash . . . I mean dot, dash it!')

Instead, I prefer the d notation due to Gottfried Leibniz, who was less of a nasty piece of work than Newton – even though Leibniz was a conniving diplomat in his day job. To show differentiation with respect to x – which simply means that x is the variable you're adjusting slightly when you do the gradient-of-the-tangent dance – you write $\frac{d}{dx}$ in front of the thing you're differentiating and mash it all up a bit: the derivative of y with respect to x is $\frac{dy}{dx}$; the derivative of $f(x)$ with respect to x is $\frac{d}{dx}f(x)$.

A second derivative would logically be $\frac{d}{dx}\left(\frac{d}{dx}f(x)\right)$, but it's more efficient to combine them together to get $\frac{d^2}{dx^2}f(x)$ or, if you're dealing with y, $\frac{d^2y}{dx^2}$.

In the exam, you're most likely to see the d-notation with variables and the prime notation with functions – that is, it's more common to see $f''(x)$ than $\frac{d^2}{dx^2}f(x)$, and you're more likely to see $\frac{dy}{dx}$ than y'.

Dealing with Powers of x

For Core 1 and Core 2, powers of x (and constants) are all you need to deal with in terms of differentiating. There's a simple recipe for differentiating a multiple of any power of x – let's say $5x^6$ – with respect to x:

1. **Multiply the coefficient by the power.**

   Here you have a coefficient of 5, which you multiply by 6 to get 30.

2. **Subtract 1 from the power and make a note of this *new power*.**

   That's 5.

3. **Write down the result from Step 1, followed by x to the new power.**

   That's $30x^5$.

There's nothing special about x. You could just as easily differentiate a power of t with respect to t or a power of z with respect to z. 'With respect to' simply means 'this letter is the variable you have to differentiate'. You see what happens when you have more than one variable in Chapter 18.

Taking care of special cases

Here are a few special cases to be aware of:

- ✔ **x to the first power:**
  - If you have to differentiate something multiplied by x, such as $-4x$, you can rewrite it as $-4x^1$ and apply the usual rule. You're left with $-4x^0$, which is -4.
  - If you have to differentiate just an x, you can rewrite it as $1x^1$, if you're very pedantically minded, and apply the rule to get $1x^0 = 1$. Alternatively, you can just remember that x differentiates to 1. It comes up often enough that it's worth knowing.

- ✔ **Constants:** If you have to differentiate a constant like 6, you can rewrite it as $6x^0$. Applying the usual rule gives you 0. Again, that's a slightly overcomplicated way of doing things: it's probably easier to remember that constants vanish when you differentiate.

You've been differentiating things like $4x$ for years: you know that the gradient of a line with the equation $y = 4x$ is 4 and that the gradient of $y = 7$, a flat line, is 0.

It's also worth mentioning negative and fractional powers, even though the recipe works the same way; it's easy to take your eye off the ball and make errors here.

For example, to differentiate $4x^{1/3}$, you would follow the rules:

1. **Multiply the coefficient by the power.**

    You get $\frac{4}{3}$.

2. **Reduce the power by 1.**

    Remembering that 1 is the same as $\frac{3}{3}$, you get $\frac{-2}{3}$, which is the new power.

3. **Write down the result.**

    You have $\frac{4}{3}x^{-2/3}$.

Similarly, to differentiate $3x^{-3}$, you would:

1. **Multiply the coefficient by the power.**

    You get –9.

2. **Reduce the power by 1.**

    $-3-1=-4$, which is the new power.

3. **Write down the result.**

    You end up with $-9x^{-4}$.

Lining up linear combinations

Differentiation has very many nice properties, but two of them are extremely useful to you at this stage: you can add derivatives up, and you can multiply them by a constant in exactly the way you'd hope.

For example, when you differentiate $af(x)$, you get $a\frac{d}{dx}f(x)$. For a more concrete example, if $y=3x^4$, $\frac{dy}{dx}=3\times\frac{d}{dx}x^4=3\times4x^3=12x^3$. If your function is multiplied by a constant, you can just ignore the constant while you're differentiating and bring it back in at the end.

Similarly, you can split up a function into things you're adding together or taking away. If you want to differentiate $f(x)+g(x)$, then you end up with $\frac{d}{dx}f(x)+\frac{d}{dx}g(x)$; you can differentiate the terms separately and add the results. For example, differentiating x^3+x^2 gives you $3x^2+2x$, the sum of the derivatives worked out individually. The same idea holds for the difference of two functions – if you differentiate $f(x)-g(x)$, you end up with $\frac{d}{dx}f(x)-\frac{d}{dx}g(x)$.

Be careful, though: if you've got functions of x multiplied or divided by each other, or – worse yet – applied to each other, you need to use at least one of the rules I show you in Chapter 17.

Differentiating Functions

Many students start Core 3 completely happy that they are masters of differentiation, able to contend with anything the examiners may dream up. Then they see the list of new functions they're meant to be able to differentiate and the new ways of combining them, and suddenly the confidence evaporates.

Take a deep breath.

For Core 3, you need to be able to differentiate all the trigonometric functions as well as exponentials and logarithms. Those are all in this chapter. You also need to be able to differentiate functions multiplied together (using the product rule), divided by each other (the quotient rule) and applied to each other (using the chain rule) – all of which are in Chapter 17.

Tackling trig functions

Sines and cosines are (despite their reputations) quite nicely behaved functions: their derivatives form a nice loop. I'll give you the mnemonic 'Drink some coffee, make some more coffee' (or D : S → C → MS → MC) and then explain what it means:

- ✔ When you're **D**ifferentiating,

    - **S**ine goes to **C**osine,

    - Which goes to **M**inus **S**ine,

    - Which goes to **M**inus **C**osine,

    - Which (back to the beginning) goes to **S**ine.

If you know your graphs for sine and cosine, you can convince yourself of these: whenever $y = \sin(x)$ is going upwards, $\cos(x)$ is positive; when it's going downwards, $\cos(x)$ is negative (see Figure 14-1). You can play similar tricks with the other relationships there.

What about the other trig functions? Your formula book has a great big table marked 'Differentiation', in which it gives you the derivatives of $\tan(x)$ and the minor trigonometric functions, along with the quotient rule, which are all you're likely to need for Core 3 and Core 4 (these may be mixed in with the derivatives of the inverse trigonometric functions and the hyperbolic functions, which you can safely ignore unless you're doing Further Maths). The derivatives you care about (in all cases, differentiating with respect to x) are these:

- The derivative of $\tan(x)$ is $\sec^2(x)$.
- The derivative of $\cot(x)$ is $-\text{cosec}^2(x)$.
- The derivative of $\sec(x)$ is $\sec(x)\tan(x)$.
- The derivative of $\text{cosec}(x)$ is $-\text{cosec}(x)\cot(x)$.

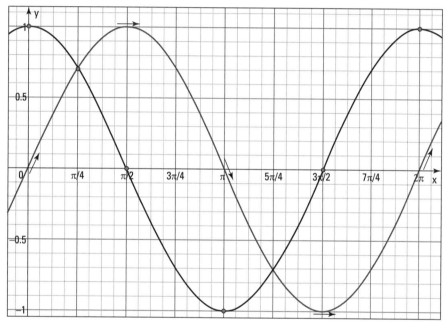

Figure 14-1:
Sine, cosine
and their
gradients.

Unless you're asked to prove these derivatives (they all succumb to the quotient rule, covered in Chapter 17, and a bit of simplification), you may look them up and use them without showing any further working; your formula book is there to be used.

Looking at logarithms and exponentials

Euler's constant, *e*, which is about 2.72, is one of the most useful numbers in maths. One of its many nice properties is that the gradient of the curve $y = e^x$ at any point is the same as the y-value: e^x is its own derivative!

Obviously, that makes differentiating it rather easy: differentiate e^x (with respect to x), and you get e^x.

The inverse function of e^x is $\ln(x)$, which isn't quite so simple to deal with; its derivative (with respect to x) is $\dfrac{1}{x}$, although it takes a bit of work to understand why. In fact, it comes from the chain rule, which you see in Chapter 17.

Finding the Gradient at a Point (and Vice Versa)

Finding the gradient of a curve requires a very simple recipe. Say you want to find the gradient of $y = x^2 + \sec(x)$ at $\left(\pi,\ \pi^2 - 1\right)$. Here's what you do:

1. Differentiate the curve.

$$\frac{dy}{dx} = 2x + \sec(x)\tan(x)$$

2. Put your given x-value into the result.

When $x = \pi$, $\dfrac{dy}{dx} = 2\pi + \sec(\pi)\tan(\pi) = 2\pi$.

That's the gradient at the given point. If you were interested in the equation of the tangent or normal there, you could find it (you know a point on the line and the gradient) – check out Chapter 15 for details on that.

If you're asked where a curve has a given gradient, on the other hand, the recipe is slightly more involved. For example, if $y = x^3 - 5x^2 - 6x + 2$ and you want to know the coordinates of the points where the gradient is –9, here's what you do:

1. Differentiate the curve.

$$\frac{dy}{dx} = 3x^2 - 10x - 6$$

2. Set the derivative equal to your target value.

$$3x^2 - 10x - 6 = -9$$

3. **Rearrange and solve the resulting equation.**

$$3x^2 - 10x + 3 = 0$$
$$(3x - 1)(x - 3) = 0$$
$$x = \frac{1}{3} \text{ or } 3$$

4. **Find the y-coordinates by substituting the x-values back into the original equation.**

When $x = \frac{1}{3}$, $y = \frac{1}{27} - \frac{5}{9} - \frac{6}{3} + 2 = -\frac{14}{27}$. When $x = 3$, $y = 27 - 45 - 18 + 2 = -34$.

5. **Give your answers clearly as points.**

The gradient is –9 at $\left(\frac{1}{3}, -\frac{14}{27}\right)$ and at $(3, -34)$.

Note that the x- and y-axes have wildly different scales, which is why the tangents look less steep in Figure 14-2 than a gradient of –9 suggests.

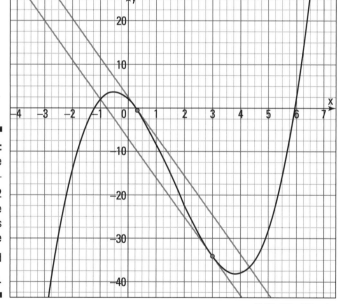

Figure 14-2:
The curve
$y = x^3 -$
$5x^2 - 6x + 2$
and the
tangents
where
$x = \frac{1}{3}$ and
$x = 3$.

Chapter 15

Touching on Tangents and Turning Points

. . . .

In This Chapter

▶ Using differentiation to find tangents and normals

▶ Identifying turning points and stationary points of inflection

▶ Maxima and minima: Taking things to extrema

▶ Finding turning points in real-world problems

. . . .

*I*t's all very well knowing *how* to differentiate – and if you don't, you should probably have a look back at Chapter 14 – but you also need to be able to apply the techniques to problems, both in geometry (Where does this curve have a specific slope? Where does it go flat?) and in real life (What's the cheapest speed to drive at? What's the biggest shape I can make out of this wire?).

In this chapter, I take you through the tangent and normal questions you may find in Core 1 (including the Really Nasty Ones), how to find turning points (largely Core 2, but it can come up in any of the modules), and the sort of artificial 'real-world' problems also normally found in Core 2.

The same thinking applies when you're given a question that involves the advanced differentiation techniques you have in Core 3 and Core 4, but those don't come up until Chapter 17!

In this chapter, it'll help if you're up to speed with the various equations of straight lines (see Chapter 10) and the techniques of differentiation (Chapter 14) – although I'll talk you through everything, it might help you to look back there if you get confused.

Finding Tangents and Normals

A *tangent* to a curve at a given point is a straight line that *just barely* touches the curve. You've seen a tangent to a circle at GCSE – if you imagine taking a bike tyre (for example) and placing a ruler against it, that's a tangent: it just touches the curve at a given point.

You may be tempted to say, 'a tangent touches the curve at only one point', but that's not strictly true: although a tangent just grazes the curve at a given point, it could then cross (or even touch) the curve again at another point far away. You see an example of that in the next section.

You define a *normal* at a given point as the straight line that's at right angles to the tangent.

Most of the work on this kind of question involves either finding the equation of a line (by way of finding its gradient and a point on it) or finding a point on a curve with a given gradient (by way of a bit of differentiation and algebra). The key in all cases is to keep track of what you know, what you're working with and what you're trying to find.

Working out the equation of a tangent

For most students, the hardest bit of a tangent question is deciphering what all of the letters mean. A typical question may ask something like this:

> The curve C has equation $y = (x-2)(x+1)(x+3)$. Point A has x-coordinate 1. Find an equation for the tangent to C at A, giving your answer in the form $ax + by + c = 0$, where a, b and c are integers.

That's a bit of an alphabet soup, isn't it? It's not quite so complicated if you break it down into bits: C is just the name of the cubic curve they've given you; A is a specific point on the curve (and they've told you the x-coordinate, generously). Meanwhile, the a, b and c in the equation of the line are integers you need to find. (A, it turns out, isn't the same thing as a.) Here's how to find the equation of the tangent:

1. Get the equation of the curve into a form you can differentiate.

In this case, you need to multiply out the brackets; you get $y = x^3 + 2x^2 - 5x - 6$.

2. Find the derivative of the curve (for general x).

If you differentiate this term by term, you get $\frac{dy}{dx} = 3x^2 + 4x - 5$.

3. Find the gradient at the given point.

Simply substitute your given x-value – here, $x = 1$ – into the derivative to get a gradient. For this one, it's 2.

4. Find the coordinates of the given point.

Here, when $x = 1$, $y = (-1)(2)(4) = -8$. The tangent passes through $(1, -8)$, which you can now identify as point A.

5. Work out the equation of the tangent line.

I'd use $(y - y_0) = m(x - x_0)$ and substitute in what I know: $(y + 8) = 2(x - 1)$; rearranging gives you $2x - y - 10 = 0$.

Getting the equation of a normal

Finding the equation of a normal is almost exactly the same as finding the equation of a tangent (see the preceding section), except that instead of using the gradient of the curve as the gradient of your line, you need to find the perpendicular gradient using the negative reciprocal rule.

A typical Core 1 question asks something like this:

A curve C has the equation $y = \dfrac{3}{x} + 2$. Find an equation for the normal to the curve at point P, where $x = -1$.

Here's your recipe:

1. Get the equation of the curve into a form you can differentiate.

In this case, you need to manipulate the power to get $y = 3x^{-1} + 2$.

2. Find the derivative of the curve (for general x).

If you differentiate this term by term, you get $\dfrac{dy}{dx} = -3x^{-2}$.

3. Find the gradient of the tangent at the given point, and then find the gradient perpendicular to it.

Here, you substitute $x = -1$ into the derivative and you find the curve has a gradient of -3 (careful with your powers, there!). The perpendicular gradient is $-\dfrac{1}{-3} = \dfrac{1}{3}$.

4. Find the coordinates of the given point.

When $x = -1$, $y = \dfrac{3}{-1} + 2 = -1$. The normal passes through $(-1, -1)$, which you can identify as point P.

5. Work out the equation of the normal line.

I'd use $(y - y_0) = m(x - x_0)$ and substitute in what I know: $(y + 1) = \dfrac{1}{3}(x + 1)$; that means $3y + 3 = x + 1$ and $x - 3y - 2 = 0$.

Working backwards: Finding points, given a gradient

Sometimes, examiners get bored of asking the same old 'find the tangent' and 'find the normal' questions and decide they need to spice up their drab lives by asking something *different*. For example, they may want you to show you can work backwards: they'll give you a value for the gradient of the curve and ask you to find the points where that's true.

As usual, the problem isn't as involved as it looks. If you want to know where the graph $y = 2x^3 + 2x^2 + 5x$ has a gradient of 7, you might try the following:

1. **Get the equation of the curve into a form you can differentiate, and differentiate it to find the derivative.**

   For this one, the derivative is $\dfrac{dy}{dx} = 6x^2 + 4x + 5$.

2. **Set the derivative equal to the given gradient; then solve.**

   $6x^2 + 4x + 5 = 7$, so $6x^2 + 4x - 2 = 0$. That factorises as $2(3x-1)(x+1) = 0$, so $x = \dfrac{1}{3}$ or $x = -1$.

3. **Find the corresponding *y*-coordinates using the original curve.**

   When $x = \dfrac{1}{3}$, $y = \dfrac{2}{27} + \dfrac{2}{9} + \dfrac{5}{3} = \dfrac{53}{27}$, and when $x = -1$, $y = -2 + 2 - 5 = -5$, so the points are $\left(\dfrac{1}{3}, \dfrac{53}{27}\right)$ and $(-1, -5)$.

Finding a tangent to a circle

Finding a tangent to a circle in Core 2 is really more of a geometry problem than a calculus problem (check out Chapter 11 for the geometric way of doing it). However, you can also work out the gradient of a tangent to a circle through a given point using implicit differentiation (which is a Core 4 topic, so don't panic if you're doing Core 2 and haven't come across it yet!).

Imagine you have everyone's favourite circle, $(x-4)^2 + (y-3)^2 = 25$, and you want the equation of the tangent through the point $(1, 7)$. Here's how it all goes down:

1. **Differentiate implicitly.**

   First expand the brackets to get $x^2 + y^2 - 8x - 6y = 0$; differentiating this implicitly with respect to x gives you $2x + 2y\dfrac{dy}{dx} - 8 - 6\dfrac{dy}{dx} = 0$.

   (Alternatively, using the chain rule on the unexpanded equation gives you $2(x-4) + 2(y-3)\dfrac{dy}{dx} = 0$, which amounts to the same thing. See Chapter 17 for the chain rule.)

2. **Substitute the known values of x and y at your given point; then solve.**

$2(-3)+2(4)\dfrac{dy}{dx}=0$, so $\dfrac{dy}{dx}=\dfrac{3}{4}$.

3. **You now have a gradient and a point, so find the equation of the line.**

$(y-y_0)=m(x-x_0)$, so $(y-7)=\dfrac{3}{4}(x-1)$, or $3x-4y+25=0$.

What else can they ask?

The examiners can ask a pretty wide range of questions after you've found tangents and normals. The good news is that most of the questions are easy enough if your coordinate geometry isn't too dicey. Some possible topics include the following:

- ✔ **Finding where a tangent or normal crosses the axes or a curve:** This is either a simple substitution problem ($x=0$ or $y=0$ on the axes) or a simultaneous-equations problem if the tangent or normal is meeting another curve.

- ✔ **Finding a point on your tangent or normal a given distance away from the tangent point:** Pythagoras (in the form of the distance formula) is your friend here.

- ✔ **Finding the area of a triangle made by tangents, normals and/or axes:** A sketch really helps with these questions. In Core 1, finding a base and a perpendicular height should always be simple; you may use Pythagoras to find some distances.

  The perpendicular distance from a point to the x-axis is just its y-coordinate (ignoring minus signs). Similarly, the distance from a point to the y-axis is its x-coordinate.

- ✔ **Finding a point on a curve where the tangent is parallel to a line or has a given gradient:** This type of problem isn't as difficult as it looks: you just need to solve for where the (general) derivate you worked out earlier is equal to the (specific) gradient at that point. Solve that equation, and you'll get values for x, some of which you'll have seen before. You want the other(s).

Sometimes, the examiners will be super-sneaky sneakers and try to slip in an $f'(x)$ instead of an $f(x)$ – in that case, they've *given* you the gradient. If they want a tangent to the curve at a given point in that case, you just need to put the point's x-coordinate into $f'(x)$ to get the gradient.

A typical triangle area question may give you a curve like $y=\dfrac{4}{x^2}$, ask you to find the tangent and the normal at the point $(2,1)$, and ask for the area of the triangle formed by the two lines and the *x*-axis. Yikes. Here's what you can do:

1. Differentiate to find the gradient function.

$$y = 4x^{-2}$$
$$\frac{dy}{dx} = -8x^{-3}$$

2. Find the equation of the tangent at the given point.

The gradient when $x = 2$ is -1, so the line is $y - 1 = -1(x - 2)$.

3. Find where the tangent line crosses the *x*-axis.

When $y = 0$, $x = 3$.

4. Find the equation of the normal at the given point.

Its gradient is 1, so the line is $y - 1 = 1(x - 2)$.

5. Find where the normal line crosses the *x*-axis.

When $y = 0$, $x = 1$.

6. Sketch your triangle.

It has corners at $(1, 0)$, $(3, 0)$ and $(2, 1)$.

7. Find the area.

The simpler way to do this is to spot that the base is 2 and the height is 1, so the area is $\frac{1}{2} bh = 1$. You may also spot that the tangent and normal are perpendicular and find the lengths of the two short sides – in this case, both of them are $\sqrt{2}$ – and say the area is (again) $\frac{1}{2} bh = \frac{1}{2} \sqrt{2} \sqrt{2} = 1$. See Figure 15-1.

Stationary Points: Turning Curves Around

A *stationary point* on a curve is anywhere the gradient is 0. In most of the situations you see at A level, this means you have a *turning point*, somewhere the curve reaches a local maximum or minimum; that is to say, the curve turns from increasing to decreasing or vice versa. *Extremum* is another word for a turning point. A question may also talk about the 'greatest possible' or 'smallest possible' – these phrases are clues that you need to look for a turning point.

There's also a kind of stationary point that's not really a turning point because the graph doesn't turn around there: it's a *stationary point of inflection*, and the section 'Classifying stationary points with the second derivative' explains all about them.

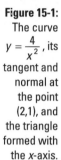

Figure 15-1: The curve $y = \dfrac{4}{x^2}$, its tangent and normal at the point (2,1), and the triangle formed with the x-axis.

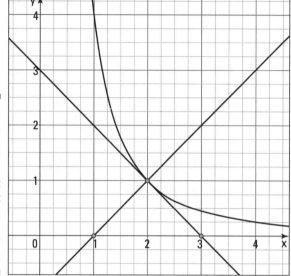

Finding stationary points with the first derivative

A curve has a *stationary point* wherever its gradient is 0, which is a great hint about how to find them. Let's take $y = x^3 - 2x^2 - 4x$ as an example.

1. **Differentiate.**

$$\frac{dy}{dx} = 3x^2 - 4x - 4$$

2. **Solve for $\dfrac{dy}{dx} = 0$.**

$$3x^2 - 4x - 4 = 0$$
$$(3x + 2)(x - 2) = 0$$

So the curve has stationary points where $x = -\dfrac{2}{3}$ or $x = 2$.

3. **Find the y-coordinates by substituting your x-values back into the equation of the curve.**

When $x = -\dfrac{2}{3}$, $y = -\dfrac{8}{27} - 2\left(\dfrac{4}{9}\right) + 4\left(\dfrac{2}{3}\right) = \dfrac{40}{27}$. And when $x = 2$, $y = 8 - 8 - 8 = -8$.

So your stationary points are at $\left(-\dfrac{2}{3}, \dfrac{40}{27}\right)$ and at $(2, -8)$.

Here, you can see that the first point is higher, so it's presumably a maximum, while the other is a minimum – and you'd be right to presume that. However, common sense isn't the way you're expected to do things in the exam; there's a mathematical way that's – for some reason – preferred.

The common-sense but nonmathematical reasoning is good only if you know the curve is *continuous* – that is, it doesn't have any holes or jumps in it. This particular curve happens to be continuous, like all polynomials, but relying on the rule of thumb is risky unless you know what you're doing.

Classifying stationary points with the second derivative

The mathematical way to classify a stationary point is to find the *second derivative* of the function at the point itself – that is, you differentiate again. Here's how to interpret the value of the second derivative:

- ✔ **Positive:** The point is a minimum (the curve has bottomed out, as in Figure 15-2a).

- ✔ **Negative:** The point is a maximum (the curve has peaked, as in Figure 15-2b).

- ✔ **Zero:** You have a stationary point of inflection (the curve goes flat and then carries on, as in Figure 15-2c).

Looking at $y = x^3 - 2x^2 - 4x$ (the example from the preceding section), here's exactly what to do:

1. **Differentiate the derivative of the curve to get the second derivative.**

    The first derivative is $\dfrac{dy}{dx} = 3x^2 - 4x - 4$, and the second derivative is

    $$\frac{d^2y}{dx^2} = 6x - 4$$

2. **Evaluate the second derivative at the stationary points.**

    The stationary points (where the first derivative equals 0) are at $\left(-\dfrac{2}{3}, \dfrac{40}{27}\right)$ and at $(2, -8)$. When $x = -\dfrac{2}{3}$, the second derivative is –8; when $x = 2$, the second derivative is 8.

3. **A positive second derivative indicates the point is a minimum; if it's negative, the point is a maximum.**

    So $\left(-\dfrac{2}{3}, \dfrac{40}{27}\right)$ is a maximum, and $(2, -8)$ is a minimum. Luckily, that corresponds with what we worked out instinctively before!

(a)

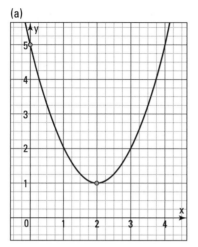

(b)

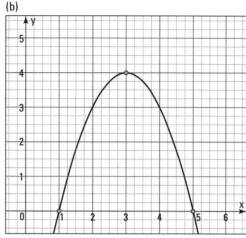

(c)

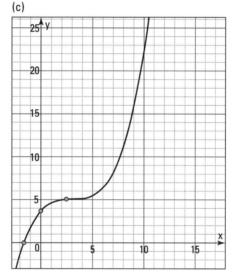

Figure 15-2:
(a) A minimum turning point on $x^2 - 4x + 5$,
(b) a maximum turning point on $y = -x^2 + 6x - 5$ and
(c) a stationary point of inflection on $y = (x-3)^3 + 5$.

© John Wiley & Sons, Inc.

'Hold on, hold on, hold on,' I hear you protesting. 'But shouldn't a maximum be positive? It's better. It's higher. It's just more *positive* than a minimum, isn't it?'

It's all those things, yes, but the second derivative doesn't measure that. The second derivative measures what's happening to the tangent line. As you go through a maximum, the tangent goes from sloping upwards (positive gradient) to flat (0 gradient) to sloping downwards (negative gradient) – the gradient is decreasing as you pass a maximum, which is why $\dfrac{d^2y}{dx^2}$, the derivative of the gradient, is negative.

Similarly, a positive second derivative gives you a minimum. In fact, you've seen this in action when drawing quadratics. A positive quadratic is a smiley-faced curve, which has a minimum. If you differentiate $y = x^2$ twice,

you get $\dfrac{d^2y}{dx^2} = 2$, which is positive for all values of x.

A positive second derivative gives you a smiley-faced minimum; a negative second derivative gives you a frowny-faced maximum. If in doubt, think about how the gradient of the tangent line at a point just to the left of the turning point differs from the gradient of the tangent line at a point just to the right.

Sketching the turning points of trig functions

You *can* find the turning points of trigonometric functions using the first- and second-derivative tests, but – at least for simple sines and cosines – you're usually as well to sketch the functions.

The sine graph has a maximum when $x = \frac{1}{2}\pi$ and every 2π after and before that (remember, any time you do calculus with trig functions, you use radians). Unless you've multiplied your sine function by a negative number, you'll have a maximum every time the argument (what's in the brackets) is $\frac{1}{2}\pi$ or any multiple of 2π more or less than that. Similarly, the sine graph has a minimum at $\frac{3}{2}\pi$ and every 2π on either side of it – so for a positive sine function, there's a minimum when the argument is $\frac{3}{2}\pi$ or any multiple of 2π more or less. If you have a negative sine function, it's the other way around!

Similarly, the cosine graph has a maximum at $x = 0$ and a minimum at $x = \pi$, and the same things appear every 2π on either side. Again, if you have a negative cosine, everything's upside down, so the maxima become minima and vice versa.

Neither sine nor cosine has a stationary point of inflection.

The graph of $y = \tan(x)$ has no stationary points: its derivative is $\sec^2(x)$, which is never 0.

As for the minor trig functions, they have turning points wherever their reciprocals do. So the turning points of $y = \operatorname{cosec}(x)$ are in the same places as those of $y = \sin(x)$, and those of $y = \sec(x)$ are in the same places as those of $y = \cos(x)$. However, because the minor trig functions are reciprocal functions, everything is upside down; what were maxima become minima and vice versa. See Figure 15-3.

The cotangent graph has no stationary points.

(a)

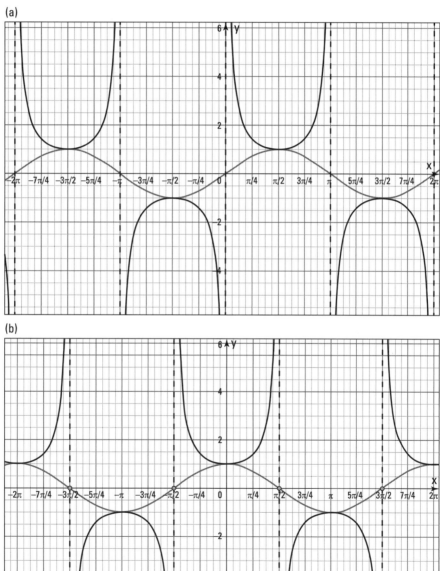

(b)

Figure 15-3: (a) The turning points of $y = \sin(x)$ and $y = \csc(x)$; (b) the turning points of $y = \cos(x)$ and $y = \sec(x)$.

Increasing and Decreasing Functions

A function is *increasing* when its graph is heading up and to the right – that is, when it has a positive gradient. Similarly, it's *decreasing* when the gradient is negative. In Figure 15-4, the function is increasing between the dashed lines, decreasing outside them and – on the lines themselves – neither increasing nor decreasing.

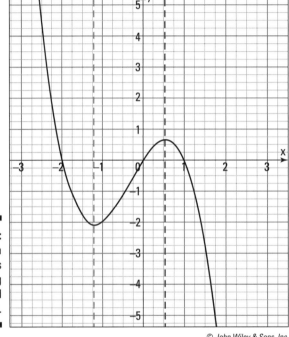

Figure 15-4:
Where a function is increasing and decreasing.

© John Wiley & Sons, Inc.

In Core 2, you may be asked to find the range of *x*-values for which a function is increasing or decreasing. Suppose you have a question like this one:

For what values of x is $f(x) = 3x^4 - 10x^3 + 9x^2 - 4$ an increasing function?

Here's how you do it:

1. **You want to know where the gradient is positive, so you need to differentiate.**

   $$f'(x) = 12x^3 - 30x^2 + 18x$$

2. **To find where the first derivative is positive, start by factorising.**

$$f'(x) = 6x(2x^2 - 5x + 3)$$
$$= 6x(2x - 3)(x - 1)$$

3. **Sketch the gradient function (see Figure 15-5) and see where it's above the x-axis.**

   This is a positive cubic that crosses the x-axis at $x = 0$, $x = 1$ and $x = \frac{3}{2}$. (See Chapter 6 if you want a refresher on sketching cubics.) You want the part between 0 and 1 as well as the part greater than $\frac{3}{2}$. The answer is that $f(x)$ is an increasing function when $0 < x < 1$ or $x > \frac{3}{2}$.

Had the question asked for where the function is decreasing, you'd need the bit of the domain where the graph of the gradient function is below the axis, and the answer would be $x < 0$ or $1 < x < \frac{3}{2}$. (If inequalities are a sticking point for you, check out Chapter 6.)

TIP

With this sort of increasing-or-decreasing-function question, the inequalities will always be strict: > rather than ≥, and < rather than ≤.

Figure 15-5: The graph of $f(x) = 3x^4 - 10x^3 + 9x^2 - 4$ and the graph of its gradient, $f'(x) = 6x(2x-3)(x-1)$.

(a) 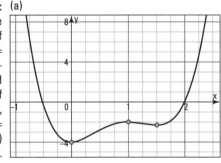   (b)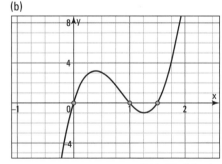

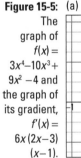

© John Wiley & Sons, Inc.

Turning Points in the 'Real World'

A typical application of turning-point problems in the real world is to find optimal solutions to problems – what's the most efficient way to use this material for packaging? What's the best time to buy or sell my shares? When will the parachutist reach terminal velocity? Unfortunately, those problems usually require a bit more maths than you have available in an exam, so you're given problems that are only occasionally realistic. (Often, there's a token attempt to shoehorn in a context, but in most cases it's just 'Here's a shape. Here are some criteria. Make it as big as you can.')

Throwing shapes

The vast bulk of the shapes questions follows the same sort of pattern:

- ✔ **You're given a shape.** Commonly, it's a cylinder or a cuboid, but sometimes it's a wire bent into an odd shape or something else entirely.

- ✔ **You're given a constraint.** For example, you may be told the volume or the surface area of a 3D shape or the perimeter or area of a 2D shape.

- ✔ **You're asked to minimise or maximise another property of the shape.** For example, given the surface area, find the greatest possible volume.

To do well with these questions, you need to be able to work out the volumes and surface areas of 3D shapes as well as the areas and perimeters of 2D shapes. Here's a quick reminder for three-dimensional shapes (r is a radius and h is a height, obviously):

- ✔ **Spheres:** volume $= \frac{4}{3}\pi r^3$; surface area $= 4\pi r^2$

- ✔ **Hemispheres:** volume $= \frac{2}{3}\pi r^3$; curved surface area $= 2\pi r^2$; base area $= \pi r^2$

- ✔ **Cylinders:** volume $= \pi r^2 h$; curved surface area $= 2\pi rh$; each end area $= \pi r^2$

- ✔ **Prisms:** volume $= Ah$; vertical surface area $= Ph$; each end area $= A$, where A is the cross-sectional area and P is the perimeter of the cross-section

- ✔ **Cones:** volume $= \frac{1}{3}\pi r^2 h$; curved surface area $= \pi r\ell$, where ℓ is the slant height; base surface area $= \pi r^2$

- ✔ **Cuboids:** volume $= hwl$; surface area $= 2(hw + wl + lh)$, where h, w and l are the height, width and length of the sides

Meanwhile, here are the formulas for two-dimensional shapes:

- ✔ **Triangles:** area $= \frac{1}{2}bh$ (where b is the base and h is the perpendicular height) or area $= \frac{1}{2}ab\sin(C)$, where a and b are the two sides adjacent to angle C

- ✔ **Circles:** area $= \pi r^2$; circumference $= 2\pi r$

- ✔ **Sectors of a circle:** area $= \frac{1}{2}\theta r^2$, where θ is the angle at the centre and r the radius; arc length $= \theta r$; perimeter $= (\theta + 2)r$

- ✔ **Rectangles:** area $= xy$, where x and y are the side lengths; perimeter $= 2(x + y)$

In some cases (for instance, with pyramids), you'll be given the relevant formulas; in others, you'll have to combine shapes or deal with a shape missing a side.

The best thing to do with complicated shapes is to break them into smaller parts – deal with each simple shape separately.

A typical maximum/minimum question may give you a cylinder with a surface area of 250 cm^2 and ask you to find the value of the radius that gives you the largest volume. Here's what you do:

1. **Write down equations for what you know.**

   Here, $V = \pi r^2 h$ and $A = 2\pi rh + 2\left(\pi r^2\right) = 250$.

2. **Eliminate one of the variables (whichever is simpler).**

   Here, you can rearrange to get h in terms of r from the area equation:

   $$2\pi rh = 250 - 2\pi r^2$$
   $$h = \frac{250 - 2\pi r^2}{2\pi r}$$
   $$h = \frac{125 - \pi r^2}{\pi r}$$

3. **Substitute into the other equation and simplify.**

   $$V = \pi r^2 \times \frac{125 - \pi r^2}{\pi r} = 125r - \pi r^3$$

4. **Now you can differentiate to find how one quantity varies as the other changes.**

   $\frac{dV}{dr} = 125 - 3\pi r^2$. This is how the volume changes as the radius changes.

5. **To find stationary point(s), solve for where this derivative is 0.**

   $0 = 125 - 3\pi r^2$, so $r = \sqrt{\frac{125}{3\pi}} \approx 3.642$ cm.

   You may notice that I use only the positive square root. That's not an error; being a distance, the radius has to be positive.

6. **Check whether you've found a maximum or a minimum by evaluating the second derivative at each stationary point.**

   Differentiate $\frac{dV}{dr}$ to get $-6\pi r$. When $r = 3.642$, this is negative, so there's a maximum at this r-value. (In other examples, you may need to do this for two or more r-values and choose the one that gives the right kind of stationary point.)

   In this case, the r-value is the answer, but . . .

7. **If the question asks for something other than the radius, substitute r back into the appropriate equation.**

If you wanted the volume here, you'd work out $125r - \pi r^3 \approx 303.5 \text{ cm}^3$ (to four significant figures) using your 'Ans' button for r as usual.

Handling 'real world' situations

Once in a while, examiners will steer away from geometry problems and ask you something from another field (sometimes literally: a farmer building a fence is quite a typical way to shoehorn a completely implausible context into a differentiation question). More often, though, it's something from the world of business. For example:

When a lorry drives at a constant speed v mph, a particular journey costs £C, where $C = \dfrac{1,000}{v} + \dfrac{2}{5}v$. Find the value of v for which the cost is a minimum. Justify that it's a minimum and find the minimum value of C.

Here's what you do:

1. **Get the equation into a form you can differentiate.**

$$C = 1,000v^{-1} + \frac{2}{5}v$$

2. **Differentiate it.**

In this case, you differentiate with respect to v:

$$\frac{dC}{dv} = -1,000v^{-2} + \frac{2}{5}$$

3. **Solve for where the first derivative is 0.**

$\dfrac{1,000}{v^2} = \dfrac{2}{5}$, so $v^2 = 2,500$ and $v = 50$ (In principle, -50 is also a solution, but common sense says reversing at 50 mph is a bad idea!)

4. **To verify that the stationary point is a minimum, differentiate the first derivative from Step 2 and check its sign at the appropriate v-value.**

If the stationary point is a minimum, the value of the second derivative there will be positive. If you have a maximum, the second derivative will be negative. The second derivative in this question is $\dfrac{d^2C}{dv^2} = 2,000v^{-3}$; when $v = 50$, the second derivative is positive, so the turning point is a minimum, as required.

5. **Put your result for v into the original equation to find the cost, C.**

$C = \dfrac{1,000}{50} + \dfrac{100}{5} = 40$. The cheapest journey costs £40, travelling at 50 mph.

Chapter 16

Integrating in Style

. . . .

. . . .

*I*ntegration is simply the reverse of differentiation. You can do two main things with integration to start with:

✔ If you're given the gradient function of a curve, you can work out the equation of the curve by integrating – although because infinitely many curves have the same gradient function, you need to know a point on the curve to specify it completely.

✔ If you're given the equation of a curve, you can work out the area underneath it by integrating between limits.

These aren't the only things you can do with integration. For example, you can solve differential equations with them, as you see in Chapter 18, although that's really a variation on the 'finding a curve given its gradient' problem I already mentioned.

Opposite Day: What Integration Is

The process of integration is exactly the opposite of the process of differentiation: whereas differentiating the equation of a curve gives you its gradient function, integrating a curve's gradient function gives you the equation of the curve (plus or minus a constant).

Integration has a more technical definition, too, which involves finding the *signed* area beneath a curve – that is to say, when the curve is above the x-axis, it counts as positive area, and when the curve is below, it counts as negative. A reasonably good way to visualise the meaning of $\int_a^b y\,dx$ (which is how you write an integral) is to think of a load of narrow vertical rectangles between a and b, each with a different height y and the same (tiny) width δx (see Figure 16-1). Depending on how many rectangles you use, adding up their areas may give you a fair approximation to the area under the curve or a good one – and no matter how many rectangles you have, it's always better to have more.

Figure 16-1: Integration effectively breaks a shape up into (a) tiny rectangles and (b) no, no, tinier than that.

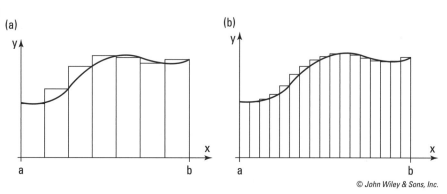

© John Wiley & Sons, Inc.

Integration is what happens as the number of rectangles goes to infinity and their width goes towards 0. As with differentiation, you don't need to go through that limiting process every time; there are handy rules you can apply to save you all that trouble.

A note on notation

To indicate an integral, you need to write down three or four parts, depending on the kind of integral you're writing:

✔ An integral symbol, a squiggle that looks like a long, curly S: $\int$

✔ The thing you're integrating, the *integrand*

✔ What you're integrating with respect to, written after a *d:* something like *dx* or *dt*

✔ *Limits,* if you have them, written with the integral symbol

For example, if you were integrating $x^2\cos(x)$ with respect to x between the limits of 0 and π, you'd write down $\int_0^\pi x^2\cos(x)\,dx$. You may get away with not writing the *dx* in Core 1 and Core 2, but it's a bad idea; firstly, it annoys the examiner, and secondly, the *dx* becomes quite important as you move through the modules. It's better to get into good habits now.

Taking Care of Constants and Powers of *x*

Broadly speaking, you need to be able to deal with two kinds of integrals:

- ✔ **Indefinite integrals**, like $\int x^2\, dx$, where there are no limits to work with
- ✔ **Definite integrals**, like $\int_1^2 x^2\, dx$, where you have values attached to the integral sign

Check out 'Looking After the Limits' later in this chapter for how to deal with the second kind; in this section, I talk about indefinite integrals.

Accounting for mystery constants

When you differentiate something, the constant terms vanish; for example, differentiating $x^2 + 7$ with respect to x gives you $2x$. Therefore, because integration is exactly the opposite of differentiation, a constant *appears* when you integrate (without limits).

Unless you've been given extra information (a point on the curve, for example, or the value of the final function for a given x), you can't tell what the constant is, so you simply have to write it down as an unknown: typically, you stick a '+ C' on the end of whatever your answer is.

Whenever you do an indefinite integral, your final answer should include a + C. Given extra information, you may be able to work out what C is.

Integrating powers of *x*

Integrating powers of x with respect to x is a little more complicated than differentiating but not very much: the recipe is almost exactly the same, except completely reversed.

For example, if you needed to integrate $8x^3$ with respect to x, here's what you'd do:

1. **Add 1 to the power and make a note of this *new power*.**

   Here, the new power is 4.

2. **Divide the coefficient by the new power.**

   If there's no written coefficient, the coefficient is 1. In this example, the coefficient is 8:

   $$8 \div 4 = 2$$

3. Write down your result from Step 2, followed by x to the new power.

$2x^4$

4. Unless you have limits, add a constant.

Your final answer is $2x^4 + C$. (See 'Looking After the Limits' later in this chapter for info on definite integrals and limits.)

As with differentiating, there's nothing special about x – you can integrate with respect to t, y, Q or *purple,* depending on what you're asked, although *purple* is an unlikely variable.

You can always check the answer to an integration question by differentiating the result – you should get back to where you started!

Finding the equation of a curve

If I had a sign that said 'beware of the elephant trap', I'd hold it up every time I saw a question that started '$\frac{dy}{dx} = \ldots$' and asked you to find a curve. I've lost count of the number of students who have blundered into the trap and said, 'It's got a $\frac{dy}{dx}$ in it, so it must be differentiation!'

Nope. If you're told what $\frac{dy}{dx}$ is, y has already been differentiated. If you want to get y back, you need to do the opposite and integrate! So if you're told that $\frac{dy}{dx} = x^3 + \frac{8}{x^2} - \frac{5}{\sqrt{x}}$ and the curve goes through $(4, 0)$, here's what you do:

1. Turn the equation into a form you can integrate.

Using your power laws, you get $\frac{dy}{dx} = x^3 + 8x^{-2} - 5x^{-1/2}$.

2. Integrate the equation with respect to x.

You get $y = \frac{1}{4}x^4 - 8x^{-1} - 10x^{1/2} + C$.

3. Substitute in the point you're given as x and y.

You get $0 = \frac{1}{4} \times 256 - 8 \times \frac{1}{4} - 10 \times 2 + C$, which simplifies to $0 = 42 + C$.

4. Solve for C.

$C = -42$

5. Write down your final answer, which is the same as Step 2 but with your value of C in its correct place.

The equation of the curve is $y = \frac{1}{4}x^4 - 8x^{-1} - 10x^{1/2} - 42$.

Marks frequently get dropped on this kind of question because students multiply by the new power instead of dividing – especially when the new power is a fraction. Make sure you know how to divide by a fraction (you multiply by the reciprocal), and it'll save you all manner of headaches.

Finding Your Way with Other Functions

When you get to Core 3 and Core 4, you're suddenly hit with the realisation that there are more things to integrate than polynomials. You also need to be able to work with the trigonometric functions, exponentials and various combinations of functions that seem impossible at first glance. (The combinations are all in Chapter 17 – I show you the others here.)

You can remember how the trigonometric functions integrate to each other using the mnemonic 'I can see my car making smoke', or I: $C \rightarrow S \rightarrow MC \rightarrow MS$. That stands for . . .

✔ When you're Integrating,

- **C**osine goes to **S**ine,

- Which goes to **M**inus **C**osine,

- Which goes to **M**inus **S**ine,

- Which – back to the start! – goes to **C**osine.

Integrating an exponential is quite straightforward, at least when the base is e: because the derivative of e^x (with respect to x) is e^x, the integral of e^x (with respect to x) is also e^x (plus a constant). Exponentials with anything else in the base (such as 2^x) are a bit trickier; see Chapter 17 for the lowdown on those.

The last integral you need to know by heart is $\int \frac{1}{x} dx = \ln(x) + C$. You're probably thinking, 'But $\frac{1}{x} = x^{-1}$, so surely you can turn it into $\frac{x^0}{0}$'. Alas, that doesn't work: you're not allowed to divide things by 0. Because the derivative of $\ln(x)$ (with respect to x) is $\frac{1}{x}$, the integral of $\frac{1}{x}$ is the natural logarithm (plus a constant).

Whenever you're doing an indefinite integral (without limits), be sure to add a constant C at the end.

Looking Things Up: Integrals from the Book

The formula book you have access to in your exam is a treasure trove of useful things to look up. Two things in particular can be life-savers: the list of compound angle formulas in trigonometry and the list of integrals that saves you from having to remember everything.

You're given the following integrals:

- ✔ The integral of $\sec^2(kx)$, which is $\frac{1}{k}\tan(kx)$, also serving as a reminder of how you deal with constants multiplied by x. (Some boards, sadly, only give that the integral of $\sec^2(x)$ is $\tan(x)$. Meanies.)

- ✔ The integral of $\tan(x)$, which is $\ln|\sec(x)|$

- ✔ The integral of $\cot(x)$, which is $\ln|\sin(x)|$

- ✔ The integral of $\operatorname{cosec}(x)$, which is $-\ln|\operatorname{cosec}(x)+\cot(x)|$ or, equivalently, $\ln\left|\tan\left(\frac{1}{2}x\right)\right|$

- ✔ The integral of $\sec(x)$, which is $\ln|\sec(x)+\tan(x)|$ or, equivalently, $\ln\left|\tan\left(\frac{1}{2}x+\frac{1}{4}\pi\right)\right|$

In all cases, there's an arbitrary constant after the integral – don't forget your $+C$!

You're also given the formula for integrating by parts, which you come across in Chapter 17.

If you're cunning, you can also use the differentiation look-up table to help integrate. Because integration is the reverse process to differentiation, if you were to integrate any of the listed derivatives in the right-hand column, you'd get the original $f(x)$ from the left-hand column.

Looking After the Limits

The only situation where you don't have to add a constant as part of your integrating is when you're using *limits*. Here's what they look like:

$$\int_{-2}^{5} x^3 - 5x\, dx$$

Avoiding the constant does come at an additional cost, though: when you're given (or can work out) limits, you need to do a few extra sums to work out your answer (which may be numerical or given in terms of one or more constants). I explain more next.

Elsewhere in this section, I show you how to use integration with limits to find the area beneath a curve, what to do when your curve crosses the *x*-axis, how to deal with areas when nice shapes or nasty curves are added or removed, and – finally – a pro tip about the order of your limits.

Simple limits

If you're given an integral with limits to evaluate, such as $\int_{-2}^{5} x^3 - 5x\,dx$, here's how to work it out:

1. **Integrate as normal, but instead of adding $+ C$ at the end, enclose the result in square brackets with the limits outside.**

$$\left[\frac{1}{4}x^4 - \frac{5}{2}x^2\right]_{-2}^{5}$$

2. **Substitute the top limit into the equation as x (or whatever variable you're working with) and write the result down.**

When $x = 5$, you get $\frac{625}{4} - \frac{125}{2} = \frac{375}{4}$.

3. **Do the same with the lower limit.**

When $x = -2$, you get $4 - 10 = -6$.

4. **Subtract the result from Step 3 from the result from Step 2.**

You get $\frac{399}{4}$.

Whenever you're integrating with limits, you always work out $(\text{answer from top limit}) - (\text{answer from bottom limit})$.

Finding an area

When you integrate between limits, what you're doing geometrically is finding the area between the curve, the *x*-axis and the vertical lines where *x* is each of the two limits – with one small caveat: area *below* the *x*-axis counts as negative area, as I explain properly in the next section.

If you know your curve *doesn't* cross the axis, finding the area between it and the axis is very easy, assuming the shape you're after is defined by vertical lines at either side. Here's how you'd find the area between the curve $y = x^2 + 3x + 4$, the x-axis and the lines $x = 1$ and $x = 4$:

1. **Write down the integral sign followed by the x-values for the vertical lines.**

   In this kind of question, the larger (rightmost) x-value goes on top. You'd write

   $$\int_1^4$$

2. **Follow this with the equation of the curve and a 'dx' to show you can cook.**

   You now have

   $$\int_1^4 x^2 + 3x + 4\, dx$$

3. **Do the integration.**

   $$\left[\frac{1}{3}x^3 + \frac{3}{2}x^2 + 4x \right]_1^4$$

4. **Evaluate this.**

   When $x = 4$, the bracket comes out to $\frac{64}{3} + 24 + 16 = \frac{184}{3}$. When $x = 1$, you get $\frac{1}{3} + \frac{3}{2} + 4 = \frac{35}{6}$. Take these away to get your area, which is $\frac{333}{6} = \frac{111}{2}$. See Figure 16-2.

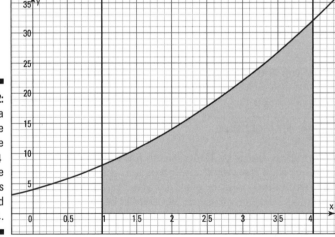

Figure 16-2: The area under the curve $y = x^2 + 3x + 4$ between the vertical lines $x = 1$ and $x = 4$.

© John Wiley & Sons, Inc.

Typically, this kind of question will ask for the *exact* area. That means leave it as a fraction. Generally, you should simplify things as far as possible, but leave fractions as fractions unless you're specifically told to give decimal answers.

Here are a couple of things to watch out for:

- ✔ **If your value works out to be negative, that's OK. It means either your area is below the *x*-axis or you have your limits back-to-front.** In either case, you need to flip the sign of your answer to get a sensible area. It's worth rechecking that your curve hasn't crossed the axis between your limits, though!

- ✔ **A variation on this kind of question asks for the area enclosed between a curve and the axis.** In this case, you need to work out the limits for yourself by solving for where the curve crosses the *x*-axis (that is to say, where $y = 0$).

Dipping below the axis

Some curves are less well-behaved, and integrating them blindly between the given limits doesn't give you the area you're interested in. For example, if you worked out $\int_{-1}^{1} x^3 - x \, dx$, you'd find you got the absurd answer of 0 for the area. (Clearly, there's *some* area between the curve and the *x*-axis!) The reasons for this answer become clear when you look at Figure 16-3, which shows the area between the curve ($y = x^3 - x$) and the *x*-axis.

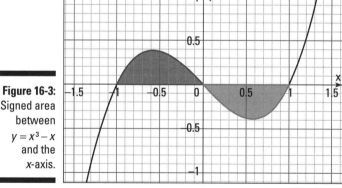

Figure 16-3: Signed area between $y = x^3 - x$ and the *x*-axis.

© John Wiley & Sons, Inc.

The shaded area on the left of the *x*-axis is above the axis and counts as positive, and the shaded area on the right is below and counts as negative. They're the same size, so they cancel each other out exactly.

So what do you do when an exam question asks you to find the area of a region that's partly above the x-axis and partly below? I'll tell you what to do, using $\int_{-1}^{1} x^3 - x \, dx$.

1. Work out the values of x where the curve crosses the x-axis.

These points split the area up into two or more regions. In this example, the curve crosses the axis at $x = 0$.

2. For each region, work out its area by integration using the appropriate limits. If the resulting area is negative (because it's below the axis), ignore the minus sign.

Here, you would work out $\int_{-1}^{0} x^3 - x \, dx = \left[\frac{1}{4}x^4 - \frac{1}{2}x^2 \right]_{-1}^{0} = 0 - \left(-\frac{1}{4} \right) = \frac{1}{4}$ for the left-hand region.

You get the same integral with different limits for the right-hand region: $\left[\frac{1}{4}x^4 - \frac{1}{2}x^2 \right]_{0}^{1} = -\frac{1}{4} - 0 = -\frac{1}{4}$. Because this part of the curve is below the axis, you need to reverse the sign to find the area, which is $\frac{1}{4}$.

3. Add up the resulting areas.

The total shaded area in this example is $\frac{1}{4} + \frac{1}{4} = \frac{1}{2}$.

The areas you need to add up aren't always the same – it's just because I've picked a nicely symmetric curve that they both happen to be a quarter in this example. Make sure you work the areas out separately.

Taking away and adding on shapes

Crossing back and forth between positive and negative areas isn't the only trick the examiners like to play on you. They're also very keen on adding shapes (usually triangles) onto what you're integrating or cutting them out.

For most people, this causes two problems: firstly, it's not always simple to spot which is the bit you need to integrate and which is the shape you need to adjust for. Secondly, it can be trickier than you'd like to find the area of the shape you need to adjust for.

Adding on a triangle

Here's an add-on triangle problem. Suppose you're given a curve, $y = x^{3/2}$, and the normal to the curve at the point $(4, 8)$, as in Figure 16-4. Your mission? To find the shaded area. Here's the brief:

1. Split the area into relatively simple areas.

In this case, you can use the area under the curve – from $(0,0)$ up to $(4, 8)$ – and the area in the triangle formed by the normal, the x-axis and the vertical line through $(4, 8)$.

2. Find the area under the curve.

By integrating, you get

$$\int_0^4 x^{3/2}\, dx = \left[\frac{2}{5}x^{5/2}\right]_0^4 = \left(\frac{64}{5}\right) - (0) = \frac{64}{5}$$

3. Find the area of the triangle.

You know something about two sides of the triangle: one side is the vertical line through $(4, 8)$, and the second lies on the x-axis. The third side of the triangle is defined by the normal to the curve at $(4, 8)$, so it makes sense to find the equation of the normal line here. Find the gradient of the curve by differentiating to get $\frac{dy}{dx} = \frac{3}{2}x^{1/2}$. When $x = 4$, the gradient is 3, so the normal gradient is $-\frac{1}{3}$. The line with that gradient through the point $(4, 8)$ is $y - 8 = -\frac{1}{3}(x - 4)$.

You want to know where the normal crosses the x-axis, so let $y = 0$, and you get $x = 28$. The base of the triangle is $28 - 4 = 24$ units wide, and its height is 8, so its area is 96.

4. Add the areas together.

You get $\frac{64}{5} + 96 = \frac{544}{5}$.

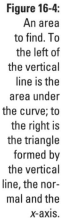
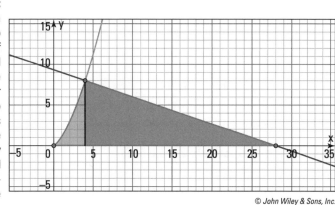

© John Wiley & Sons, Inc.

Taking off a triangle

When the area beneath your curve has a triangle gouged out of it (or other shape, but it's normally a triangle), the steps are fairly similar to the steps for adding a triangle (see the preceding section). But this time, you're using the idea of negative space to take away a shape rather than add it on.

REMEMBER

Think of a shape that's been cut away as a negative area, or an area taken away from a bigger area.

For example, say you have the curve $y = x^{3/2}$ and the tangent to the curve at the point $(4, 8)$, as in Figure 16-5. You need to find the shaded area. The only difference between this problem and the preceding one is that now you're dealing with a tangent rather than a normal. (**Note:** Integrating gives the area of the 'triangle' formed by the curve from $(0, 0)$ to $(4, 8)$, the x-axis and the vertical line; the real triangle formed by the vertical line, the x-axis and the tangent is subtracted to leave the sliver you're interested in.)

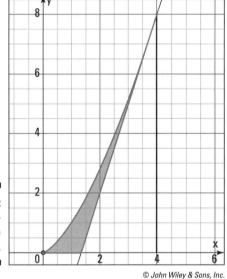

Figure 16-5:
A negative-space shape.

© John Wiley & Sons, Inc.

Here are your instructions:

1. **Split the area into relatively simple areas.**

   In this case, you can use the area under the curve from $(0, 0)$ to $(4, 8)$. The second area is the *negative* area of the triangle that's cut out of it.

2. **Find the area under the curve.**

   By integrating, you get

   $$\int_0^4 x^{3/2}\, dx = \left[\frac{2}{5} x^{5/2}\right]_0^4 = \left(\frac{64}{5}\right) - (0) = \frac{64}{5}$$

   You may have a sense of déjà vu. You may have a sense of déjà vu.

3. **Find the area of the triangle.**

   You know the triangle is defined by the x-axis, the vertical line through $(4, 8)$ and the tangent to the curve, so it makes sense to find the equation of the tangent at that point. Find the gradient of the curve by differentiating to get $\frac{dy}{dx} = \frac{3}{2}x^{1/2}$. When $x = 4$, the gradient is 3. The line with that gradient through the point $(4, 8)$ is $(y - 8) = 3(x - 4)$.

   You want to know where the tangent crosses the x-axis. Let $y = 0$, and you get $x = \frac{4}{3}$. The base of the triangle is $4 - \frac{4}{3} = \frac{8}{3}$ and its height is 8, so its area is $\frac{32}{3}$.

4. **Take the areas away.**

   You get $\frac{64}{5} - \frac{32}{3} = \frac{32}{15}$.

Taking away curves

Finding the area between curves is a relatively simple assignment. For example, say you need to find the area between $y = 3e^{2x} - 2$ and $y = 9e^x - 8$, as shown in Figure 16-6.

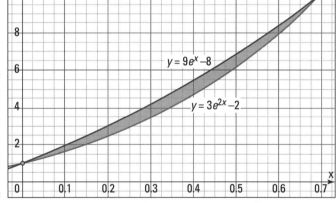

Figure 16-6:
The area between two curves, $y = 3e^{2x} - 2$ and $y = 9e^x - 8$.

Your approach here is to work out the limits you want to integrate between by finding where the curves intersect. Then you find the area beneath each curve so you can find the difference. Here's the full recipe:

1. Find the *x*-coordinates of the points where the curves intersect.

$3e^{2x} - 2 = 9e^x - 8$ is a disguised quadratic; let $u = e^x$ and rearrange to get $3u^2 - 9u + 6 = 0$, which factorises as $3(u-1)(u-2) = 0$. That means $e^x = 1$ or $e^x = 2$, and $x = 0$ or $x = \ln(2)$.

2. Integrate one curve using your solutions to Step 1.

$$\int_0^{\ln(2)} 3e^{2x} - 2\,dx = \left[\frac{3}{2} e^{2x} - 2x \right]_0^{\ln(2)}$$
$$= \left(6 - 2\ln(2) \right) - \left(\frac{3}{2} \right)$$
$$= \frac{9}{2} - 2\ln(2)$$

3. Integrate the other curve using the same limits.

$$\int_0^{\ln(2)} 9e^x - 8\,dx = \left[9e^x - 8x \right]_0^{\ln(2)}$$
$$= 9 - 8\ln(2)$$

4. Find the difference between two areas.

You get $\left(9 - 8\ln(2) \right) - \left(\frac{9}{2} - 2\ln(2) \right) = \frac{9}{2} - 6\ln(2)$.

If you're in any doubt about which area to take away from which, you have these three options, roughly in the order I'd recommend them:

✔ Draw a sketch to work out which curve is above the other (you take the area beneath the lower curve away from the area under the upper curve).

✔ Evaluate the curves at a point somewhere between the limits to check which *y*-value is larger (that curve will be the upper curve).

✔ Guess and then check whether your final area is positive (if not, you have it the wrong way around).

If you need to find the area of a region between a curve, some horizontal lines and the *y*-axis, like the region in Figure 16-7, there are several approaches; I show you a couple here. Suppose you need the area between the curve $y = 1 + \sqrt{x}$, the *y*-axis and the line $y = 5$.

I would approach this problem with subtraction. There are two areas of interest: a) the rectangle formed by the axes, the horizontal line and the vertical line through the point where the line and curve cross and b) the area under the curve. Taking the area under the curve away from the rectangle leaves the area you want.

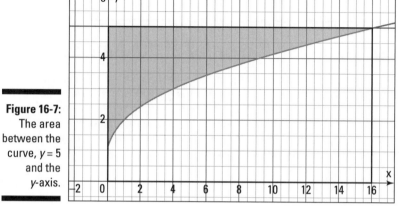

Figure 16-7:
The area between the curve, $y = 5$ and the y-axis.

1. **Find the limits for integration.**

   The left-hand end of the area is at $x = 0$. The right-hand end is where the curve crosses the horizontal line, where $5 = 1 + \sqrt{x}$, so $x = 16$.

2. **Find the area beneath the upper curve.**

   In this case, the upper curve is a line, so the area beneath it is a rectangle; its area is $16 \times 5 = 80$.

3. **Find the area beneath the lower curve.**

   You need to work out

   $$\int_0^{16} 1 + x^{1/2}\, dx = \left[x + \frac{2}{3} x^{3/2} \right]_0^{16} = \left(16 + \frac{128}{3} \right) - (0) = \frac{176}{3}$$

4. **Subtract the result from Step 3 from the answer from Step 2.**

   $$80 - \frac{176}{3} = \frac{64}{3}$$

There's another approach, which is usually overkill, but it's worth mentioning all the same. If you can write the curve as $x = f(y)$, you can integrate with respect to y instead of x. Your limits will be in terms of y rather than x, but otherwise everything proceeds as normal. In this example, you'd rearrange the equation as $x = y^2 - 2y + 1$. Integrating this between the y-limits of 1 and 5 gives you

$$\int_1^5 y^2 - 2y + 1\, dy = \left[\frac{1}{3} y^3 - y^2 + y \right]_1^5$$
$$= \left(\frac{125}{3} - 25 + 5 \right) - \left(\frac{1}{3} - 1 + 1 \right)$$
$$= \frac{64}{3}$$

as before.

TIP

Pro tip: Changing the order of the limits

There's a nice trick you can use if your integral is multiplied by a nasty minus sign, if your limits are 'the wrong way around' (that is, with the top number smaller than the bottom), or both. It's this: if you change the order of the limits, you get the same answer with the opposite sign. For example, $\int_0^4 3x^2\,dx = \left[x^3\right]_0^4 = 64 - 0 = 64$.

Meanwhile, $\int_4^0 3x^2\,dx = \left[x^3\right]_4^0 = 0 - 64 = -64$.

This means that if you've got a minus sign that you expect is going to get awkward – especially if you're doing integration by parts (see Chapter 17) – you can simply switch the limits and turn it into a plus that's less confusing to deal with!

To Infinity and Beyond

In certain boards, you may be asked to deal with integrals that look like this:

$$\int_4^\infty 4x^{-3}\,dx$$

You may recognise the sideways 8 at the top as the symbol for 'infinity'. In this section, I show you two ways to deal with it: the proper, rigorous way that should get all the marks, and a hand-wavy way that will make mathematicians recoil with horror and you definitely shouldn't use. Definitely not, even if it should normally get the right answer and leave you with most of the marks.

The proper way to do the problem goes like this:

1. **Replace the infinity symbol with *N* – think of *N* as just a very big number.**

   Your integral is now $\int_4^N 4x^{-3}\,dx$.

2. **Work out the integral in terms of *N*.**

   You get

   $$\left[-2x^{-2}\right]_4^N = -\frac{2}{N^2} - \left(-\frac{2}{4^2}\right) = -\frac{2}{N^2} + \frac{1}{8}$$

3. Write 'as $N \to \infty$, $-\dfrac{2}{N^2} \to 0$'.

For a different problem, your fraction will obviously be different, although it's still very likely to go to 0.

This is what the ∞ in the limit really means: it's a shorthand for 'What happens to the integral as N gets bigger and bigger without bound?' Just like the sum of a geometric series (with an appropriately small common ratio) approaches a fixed value as you increase the number of terms, so some integrals approach a fixed value as you increase the upper limit. (A consequence: sometimes you need only a finite amount of paint to colour in the area below an infinitely long line.)

4. Finish off by giving your final answer.

$$\frac{1}{8}$$

This method is called using a *limiting process*, and it avoids any potential difficulties with infinity.

At A level, you will only ever see fractions that become very small as N gets very big, so the thing with the arrows in Step 3 will stand you in good stead. In other bits of maths, you may find that the Ns don't go away; that would mean that the integral doesn't exist.

The hand-wavy way – remember, don't do it like this, even if I'm in the other room looking the other way – is to suppose that anything divided by infinity is 0. This way of thinking about things is a workable rule of thumb, but it can lead to problems later in maths – you can't treat infinity as if it's a number, or you get into some very nasty paradoxes (and *very* stern looks from proper mathematicians).

Working out the integral the hand-wavy way, you get

$$\left[-2x^{-2} \right]_4^\infty = \left(-\frac{2}{\infty^2} \right) - \left(-\frac{2}{4^2} \right) = 0 + \frac{1}{8}$$

This is the correct final answer, but $\dfrac{2}{\infty^2}$ is extremely bad form mathematically. Infinity isn't a number, and you can't divide by it any more than you can divide by 0.

The wavy method should, however, get you most of the marks available. But if anyone asks, it wasn't me who told you about it, and I definitely didn't recommend it to you, understand?

Chapter 17

When to Reach for the Rules

Core 3, for most exam boards, is substantially about whether you can differentiate well. Core 4, on the other hand, is mainly integration – only, it's not (just) the powers of x you know and love. Now you're dealing with functions like $\sin(x)$ and $\ln(x)$ and (even more distressingly) combinations of them.

However, this isn't 'Nam – there are *rules*: three main ones for differentiating (the chain, product and quotient rules) and two main ones for integrating (substitution and parts). There are other techniques, too, for dealing with nasty trigonometric integrals, but those are in Chapter 18.

Calculus with Linear Expressions

A *linear expression* is anything you can write as $ax + b$ (where a and b are constants) because you could make a straight line by setting y equal to it. Here are some examples of linear expressions:

✔ $3x + 4$

✔ x (you can write it as $1x + 0$)

✔ $9 - x$ (you can write it as $-1x + 9$)

✔ $\pi - 3x\sqrt{2}$ (you can write it as $\left(-3\sqrt{2}\right)x + \pi$

✔ $\frac{4x-5}{3}$ (you can write it as $\frac{4}{3}x - \frac{5}{3}$)

I mention linear expressions because you *very* frequently need to integrate or differentiate functions with them in brackets – things like $(3x+4)^5$ or $\sin(2x-3)$ or $e^{(7/2)-\pi x}$, which I like to call *nested linear expressions*. Because they're common, I'll take a moment to go through how to integrate and differentiate, and I'll explain why they work the way they do.

Differentiating nested linear expressions

You know how to differentiate $\sin(x)$ – at least, if you've read Chapter 14, you should; you get $\cos(x)$. But suppose x has a few friends inside the bracket. What about differentiating $\sin(2x-3)$, for example?

The short answer? You differentiate as normal, but you multiply by the number in front of the x. You get $\frac{d}{dx}\big(\sin(2x-3)\big) = 2\cos(2x-3)$. Similarly, if you differentiate e^{-4x}, you get $-4e^{-4x}$. The thing in the brackets (or implied brackets) doesn't change.

Why is this? It's because of the chain rule, which you see a bit later in this chapter. However, there's an intuitive reason as well: you remember Bad Guy x from Chapter 10? The graph of $y = f(ax+b)$ is translated left by b (which you don't much care about) and squished towards the axis by a factor of a, effectively making the graph a times steeper. That's why you multiply the gradient by a.

In case you want a rule you can write down for this, here it is:
$\frac{d}{dx}\big(f(ax+b)\big) = af'(ax+b).$

Integrating nested linear expressions

Hopefully, you're happy integrating things like $\sec^2(x)$ with respect to x – Chapter 16 is where to go if not; you should get $\tan(x)+C$. But what if a linear expression is in the bracket, as in $\sec^2(4-3x)$?

Well, integration is the opposite of differentiation, so it shouldn't be too much of a surprise that when integrating, you *divide* by the number in front of the x to get $-\frac{1}{3}\tan(4-3x)+C$. Similarly, the integral of $(2x+3)^5$ is $\frac{1}{12}(2x+3)^6 +C$, because you divide by the new power (6) and by the 2 in front of the x. Again, what's in the brackets doesn't change.

Why does this work? Again, the graph of $y = f(ax+b)$ is translated left by b and squished towards the axis by a factor of a. Not only does squishing the graph multiply the gradient by a factor of a, but it also reduces the graph's area by a factor of a.

Again, for the benefit of those who like to write down a rule, if
$\int f(x)dx = F(x)+C$, then $\int f(ax+b)dx = \frac{1}{a}F(ax+b)+C$.

Differentiating with the Chain, Product and Quotient Rules

It's not enough to know the derivatives of the main functions introduced in Chapter 14 (powers of x, trig functions, natural logs and the like) – you need to be able to differentiate combinations of them, too. There are three combinations you care about in Core 3:

- ✔ **Function of a function:** For example, $f(x) = \cos\left(\ln\left(x^2+3\right)\right)$ is a function wrapped inside a function wrapped inside a third function! To differentiate this sort of case, you use the chain rule.
- ✔ **Multiplied functions:** If $f(x) = x^2 \tan(x)$, you need to use the product rule to differentiate – *product* is another word for 'multiplication'.
- ✔ **Divided functions:** If $f(x) = \dfrac{e^x}{3\sin(x)}$, you use the quotient rule for differentiation – *quotient* means 'division'.

You may need to use these rules one after the other – if $f(x) = \dfrac{3xe^{x^2}}{\tan(x)}$, you need all three!

Chain rule

The chain rule is for differentiation questions that look like a nightmare, such as 'Let $y = \cos\left(\ln\left(x^2+3\right)\right)$. Find $\dfrac{dy}{dx}$.' (This one is probably tougher than you can expect to see in an exam, but it illustrates what's going on.)

The trick to the chain rule is to break the function down into little bits that you *can* differentiate – and then multiply them all together at the end. Here's a recipe:

1. **Take what's in the first set of brackets and call it u.**

    Here, you get $y = \cos(u)$, where $u = \ln\left(x^2+3\right)$.

2. **Repeat this (using other letters) for each remaining set of brackets.**

   You have $u = \ln(v)$, where $v = x^2 + 3$. I'd lay this out as follows:

   $$y = \cos(u) \qquad u = \ln(v) \qquad v = x^2 + 3$$

3. **Differentiate y, u and v (and others, if you have them), as appropriate.**

   $$\frac{dy}{du} = -\sin(u) \qquad \frac{du}{dv} = \frac{1}{v} \qquad \frac{dv}{dx} = 2x$$

4. **Multiplying these together gives you $\frac{dy}{dx}$.**

   $$\frac{dy}{dx} = -\sin(u)\left(\frac{1}{v}\right)(2x)$$

5. **You're not quite done: you need to turn all those made-up letters back into xs and tidy up.**

   $$\frac{dy}{dx} = -\sin\left(\ln\left(x^2 + 3\right)\right)\left(\frac{1}{x^2 + 3}\right)(2x)$$
   $$= -\frac{2x}{x^2 + 3}\sin\left(\ln\left(x^2 + 3\right)\right)$$

If you like to write down general forms of rules, here's the chain rule: $\frac{dy}{dx} = \frac{dy}{du} \times \frac{du}{dv} \times \frac{dv}{dx}$. This extends for as many intermediate steps as you need. (Be careful: you're not technically 'cancelling' dus and dvs, even if it looks that way.)

Linear expressions revisited

The chain rule also explains why the rule for linear expressions works. If you need to differentiate $y = \tan(2x + 3)$, here's the chain rule:

1. **Rewrite.**

   $y = \tan(u)$ and $u = 2x + 3$.

2. **Differentiate both.**

   You get $\frac{dy}{du} = \sec^2(u)$ and $\frac{du}{dx} = 2$.

3. **Multiply them together.**

   $$\frac{dy}{dx} = \frac{dy}{du} \times \frac{du}{dx} = 2\sec^2(u)$$

4. Replace *u*.

$\frac{dy}{dx} = 2\sec^2(2x+3)$, as you (hopefully) expected.

Differentiating complicated logarithms

There's nothing substantially new in this section, but differentiating complicated logs is a notorious stumbling block – and something I always used to struggle with. Your first example is a moderately complicated one: $y = \ln(4x^3 + 5)$ – but you're going to approach it just the same way as the others.

1. Rewrite.

$y = \ln(u)$ and $u = 4x^3 + 5$.

2. Differentiate both.

$\frac{dy}{du} = \frac{1}{u}$ and $\frac{du}{dx} = 12x^2$.

3. Multiply them together.

$$\frac{dy}{dx} = \frac{dy}{du} \times \frac{du}{dx} = \frac{1}{u} \times 12x^2 = \frac{12x^2}{u}$$

4. Replace *u*.

$$\frac{dy}{dx} = \frac{12x^2}{4x^3 + 5}$$

Oddly, the one I always found trickiest was actually simpler: differentiating $y = \ln(3x)$. I'd always jump in and say, 'Oh! It's $\frac{1}{3x}$.' Wrong. Now, I do it more carefully: rewrite it as $y = \ln(u)$ and $u = 3x$; differentiate both to get $\frac{dy}{du} = \frac{1}{u}$ and $\frac{du}{dx} = 3$; multiply them together to get $\frac{dy}{dx} = \frac{dy}{du} \times \frac{du}{dx} = \frac{1}{u} \times 3 = \frac{3}{u}$; and replace *u* to get $\frac{3}{3x} = \frac{1}{x}$.

But hang on ... isn't that the derivative of $\ln(x)$? Yes, yes it is. You can also write $\ln(3x)$ as $\ln(x) + \ln(3)$; when you differentiate it, the $\ln(3)$ (being a constant) vanishes.

Product rule

You use the product rule when you have to differentiate an expression that's made of two things multiplied together. For example, you'd use the product rule to differentiate $x^2(3+x)^3$, $x\cos(x)$ or $e^x\cos(2x)$.

If you call one of your factors u and the other one v, the derivative of the whole thing is $u\dfrac{dv}{dx} + v\dfrac{du}{dx}$, or (equivalently) $u\dfrac{dv}{dx} + \dfrac{du}{dx}v$ – which is something you need to know by heart (some students find the second form easier to memorise, as there's a nice symmetry to it). I'd work through $e^x\cos(2x)$ like this (follow along with Figure 17-1):

1. **Write down u and v.**

   $u = e^x$ and $v = \cos(2x)$.

2. **Differentiate and write the answers underneath.**

   Differentiating gives you $\dfrac{du}{dx} = e^x$ and $\dfrac{dv}{dx} = -2\sin(2x)$.

3. **Draw a cross (as in Figure 17-1) and multiply the appropriate pairs.**

   $u\dfrac{dv}{dx} = -2e^x\sin(2x)$ and $v\dfrac{du}{dx} = e^x\cos(2x)$.

4. **Add these up and simplify.**

   You get $e^x\big(\cos(2x) - 2\sin(2x)\big)$.

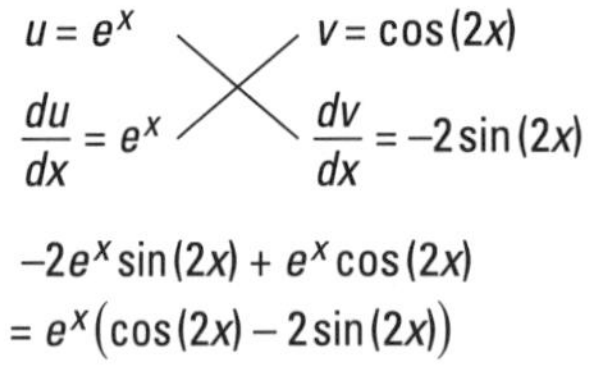

Figure 17-1:
Product rule differentiation.

$$u = e^x \qquad v = \cos(2x)$$
$$\frac{du}{dx} = e^x \qquad \frac{dv}{dx} = -2\sin(2x)$$
$$-2e^x\sin(2x) + e^x\cos(2x)$$
$$= e^x\big(\cos(2x) - 2\sin(2x)\big)$$

© John Wiley & Sons, Inc.

Quotient rule

The quotient rule is the only one of the differentiation rules you're given in the book – although it's given in a form that's not exactly obvious. The name gives you a clue about when you use it: *quotient* is another word for division (you see quotients and remainders in Chapter 5). You use the quotient rule when you have a fraction you want to differentiate – for example, $y = \dfrac{\cos(x)}{x^2}$.

If you define u as the top of your fraction and v as the bottom, then the quotient rule says the derivative of the fraction is

$$\frac{v\dfrac{du}{dx} - u\dfrac{dv}{dx}}{v^2}$$

Why is the quotient rule this unholy mess? You can work it out, if you're patient, using the product rule and the chain rule. Instead of writing the thing you're differentiating as $\frac{u}{v}$, write it as uv^{-1}. Split this up as $U = u$ and $V = v^{-1}$; differentiating these gives you $\frac{dU}{dx} = \frac{du}{dx}$ and $\frac{dV}{dx} = -1v^{-2}\frac{dv}{dx}$. Applying the product rule, you get $U\frac{dV}{dx} + V\frac{dU}{dx} = \frac{\left(-u\frac{dv}{dx}\right)}{v^2} + \left(\frac{1}{v}\right)\frac{du}{dx}$. Turning this into a single fraction with v^2 on the bottom gives you the quotient rule you know and loathe.

The most sensible way I know to apply the quotient rule is to make a sort of table with u and v in the top row and their derivatives in the row below. To remember where to start, I tend to draw a memorable fish (as in Figure 17-2) starting in the top right and squiggling around.

Here's how you'd differentiate $y = \frac{\cos(x)}{x^2}$:

1. **Write $u = \left(\text{whatever's on the top}\right)$ and $v = \left(\text{whatever's on the bottom}\right)$ side by side, as in Figure 17-2a.**

2. **Differentiate u and write $\frac{du}{dx}$ under u; differentiate v and write $\frac{dv}{dx}$ under the v.**

   Here, you have $\frac{du}{dx} = -\sin(x)$ and $\frac{dv}{dx} = 2x$.

3. **Draw the memorable fish to remind you to work out $v\frac{du}{dx}$ and subtract $u\frac{dv}{dx}$.**

   The top of your answer will be $-x^2\sin(x) - 2x\cos(x)$.

4. **The bottom, meanwhile, is v^2.**

   Here, that's x^4, making the answer $\dfrac{-x^2\sin(x) - 2x\cos(x)}{x^4}$,

   or $-\dfrac{x\sin(x) + 2\cos(x)}{x^3}$.

Figure 17-2: Quotient rule differentiation.

(a) $u = \cos(x)$ $v = x^2$
$\frac{du}{dx} = -\sin(x)$ $\frac{dv}{dx} = 2x$

(b) $\dfrac{-x^2\sin(x) - 2x\cos(x)}{x^4}$

$= -\dfrac{x\sin(x) + 2\cos(x)}{x^3}$

Inverting $\dfrac{dy}{dx}$

Questions where you're given x in terms of y (for example, $x = \operatorname{cosec}^2(4y)$ for $x > 1$ and $\frac{\pi}{8} < y < \frac{\pi}{4}$) and you need to find $\frac{dy}{dx}$ in terms of x are, in my head, classified as 'oh, you just...' questions. 'Oh you just...' questions are the kind of questions where you say, 'Oh, you just differentiate and flip it upside down' (or whatever the supposedly simple one-step solution is). Technically speaking, that's 'right', but then technically all you have to do in a football match is stick the round thing through those goalpost-majiggers and you're all good. In both cases, that's the essence – but the details are a bit more complicated than they seem at first.

In this case, getting $\frac{dx}{dy}$ (note that this is the 'other way up' from what you're used to; you're differentiating with respect to y rather than x) is fairly straightforward: writing $x = \left(\sin(4y)\right)^{-2}$ and using the chain rule gives you $\dfrac{dx}{dy} = -8\cos(4y)\left(\sin(4y)\right)^{-3} = -\dfrac{8\cos(4y)}{\sin^3(4y)}.$

(In the domain you're interested in, $\cos(4y)$ is negative and $\sin^3(4y)$ is positive, so $\frac{dx}{dy}$ is positive. This will come in useful later.)

Now, it certainly *looks* like you can turn $\frac{dx}{dy}$ into $\frac{dy}{dx}$ by simply flipping the fraction upside down – and if you're a physicist or some other lesser mortal, you may get away with that. But here in maths-land, I'm afraid that's not quite it. You see, $\frac{dx}{dy}$ is not a fraction; it's notation. It does turn out that $\dfrac{1}{\left(\dfrac{dx}{dy}\right)} = \dfrac{dy}{dx},$ but this comes from the chain rule and not from fraction rules.

The first part is easy enough. The tricky bit is getting everything back in terms of x. (This is really an exercise for Chapter 12, but indulge me.)

It's tricky to give precise instructions for this sort of question, as there are so many variants, but a decent strategy is to simplify in bits: keep trying to turn what you have into something you know about.

For example, you know $x = \operatorname{cosec}^2(4y)$, so how could you get some cosecs? You could write $\sin^3(4y)$ as $\dfrac{\sin(4y)}{\operatorname{cosec}^2(4y)}$, which is $\dfrac{\sin(4y)}{x}$. That makes the fraction $\dfrac{dy}{dx} = -\dfrac{1}{8} \times \dfrac{\sin(4y)}{x} \times \dfrac{1}{\cos(4y)} = -\dfrac{1}{8x}\tan(4y)$. Now, how do you get a tan, or possibly a cot? Well, there's a link between cosec and cot:

$1 + \cot^2(4y) \equiv \operatorname{cosec}^2(4y)$. That means $\cot(4y) \equiv \pm\sqrt{\operatorname{cosec}^2(4y) - 1} = \pm\sqrt{x-1}$, so $\tan(4y) = \pm\dfrac{1}{\sqrt{x-1}}$.

You've got $\dfrac{dy}{dx} = -\dfrac{1}{8x}\tan(4y) = \pm\dfrac{1}{8x\sqrt{x-1}}$, only you know (because $\dfrac{dx}{dy}$ is positive in the domain of interest) that it must be positive, so

$$\frac{dy}{dx} = \frac{1}{8x\sqrt{x-1}}.$$

You'd likely be asked to show the final answer rather than find it from scratch – it's nice to have a target to work towards!

Linked derivatives

Another real-world application of derivatives that comes up in Core 4 is the idea of *linked derivatives*. This is what happens when you have several variables with equations linking them and, given some of their derivatives, you work out the others.

One thing to flag up before you start: if you see the word *rate*, it almost certainly means 'derivative with respect to time'. So if you see 'the area is increasing at a rate of 7 cm^2 per second', it means $\dfrac{dA}{dt} = 7$. You can also use the unit to help you: area is measured in cm^2 and time is measured in seconds, so a unit of cm$^2 \cdot$s^{-1} is $\dfrac{\text{area}}{\text{time}}$, meaning $\dfrac{dA}{dt}$ makes sense.

For example, suppose you have a cuboid (as in Figure 17-3) with sides of x, $2x$ and $4x$, and you know its volume is increasing at a rate of 0.4 cm$^3 \cdot$s^{-1}. Typically, the question would ask something like 'At what rate is x increasing when the volume is 64 cm^3?'

Here's how you'd do it:

1. **Write down what you want and what you've got.**

   You want $\dfrac{dx}{dt}$, and you have $\dfrac{dV}{dt} = 0.4$ cm$^3 \cdot$s^{-1} as well as your relationship $V = 8x^3$.

2. **Contemplate the derivatives.**

   Your relationship is all Vs and xs, and you're interested in t-derivatives. No matter: you can differentiate with respect to t!

3. **Differentiate using the chain rule.**

   Differentiating $V = 8x^3$ gives you $\dfrac{dV}{dt} = 24x^2\dfrac{dx}{dt}$. You can rearrange to get $\dfrac{dx}{dt} = \dfrac{1}{24x^2}\dfrac{dV}{dt}$.

4. **You're not done yet: you need to work out x when V is 64 using the volume equation.**

$64 = 8x^3$, so $x = 2$.

5. **Put known values for x and $\dfrac{dV}{dt}$ into the derivative to get your answer.**

$$\frac{dx}{dt} = \frac{0.4}{24 \times 2^2} = \frac{1}{240} \ \text{cm} \cdot \text{s}^{-1}$$

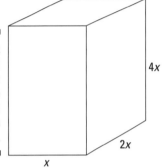

Figure 17-3:
A cuboid with given sides (of the kind you might see in an exam).

$4x$

$2x$

x

© John Wiley & Sons, Inc.

Putting the rules together

A typical Core 3 question may ask something like this:

A curve C has the equation $y = \sqrt{\cos\left(3x - \frac{1}{6}\pi\right)}$. Find the normal to the curve at the point where $x = \frac{\pi}{6}$.

Here's what you'd do:

1. **Turn the curve into something you can differentiate.**

$$y = \left(\cos\left(3x - \frac{1}{6}\pi\right)\right)^{1/2}$$

2. **Differentiate using the appropriate rules.**

Here, it's the chain rule all the way down: set it up as $y = u^{1/2}$, $u = \cos(v)$ and $v = 3x - \frac{1}{6}\pi$. Then $\dfrac{dy}{dx} = \dfrac{dy}{du} \times \dfrac{du}{dv} \times \dfrac{dv}{dx}$, which works out to be

$$\tfrac{1}{2}u^{-1/2}\times\left(-\sin(v)\right)\times 3$$

$$=-\frac{3}{2}\left(\cos\left(3x-\tfrac{1}{6}\pi\right)\right)^{-1/2}\sin\left(3x-\tfrac{1}{6}\pi\right)$$

You did that in your head, right?

3. Substitute in your x-value to find the gradient of the curve.

Here, $x=\tfrac{1}{6}\pi$, so $3x-\tfrac{1}{6}\pi=\tfrac{1}{3}\pi$.

Next, $\cos\left(\tfrac{1}{3}\pi\right)=\tfrac{1}{2}$ and $\sin\left(\tfrac{1}{3}\pi\right)=\dfrac{\sqrt{3}}{2}$.

The gradient is $-\dfrac{3}{2}\left(\dfrac{\sqrt{3}}{2}\right)\left(\dfrac{1}{2}\right)^{-1/2}=-\dfrac{3\sqrt{6}}{4}$.

4. If you want the normal (which you do here), find the negative reciprocal.

$$m=\frac{4}{3\sqrt{6}}$$

You could rationalise that if you wanted, but you don't need to here.

5. Find the y-coordinate from the original equation (the one with a y in it).

$y=\sqrt{\cos\left(\tfrac{1}{3}\pi\right)}=\dfrac{1}{\sqrt{2}}$. Your point is at $\left(\dfrac{\pi}{6},\dfrac{1}{\sqrt{2}}\right)$.

6. Now you have all you need to find the equation of the line.

$$y-\frac{1}{\sqrt{2}}=\frac{4}{3\sqrt{6}}\left(x-\tfrac{1}{6}\pi\right)$$

In principle, that's a perfectly good equation of a line, and unless the question asks you for a specific form, you can leave it there. If you're asked to tidy it up, though, it can be rearranged to $12x-9\sqrt{6}y+9\sqrt{3}-2\pi=0$.

Leave things in exact form wherever you can – πs and square roots are perfectly OK in your answer. Fractions are also fine, although I've managed to get rid of them here as a matter of style.

Exponentials: Tricky Derivatives

There are some things that get thrown at you more out of cruelty than anything else: for example, differentiating things like a^{x} (where a is a constant) and x^{x}. There's a booby-trap here: almost everyone, the first time they're asked to differentiate a^{x} (with respect to x), says 'it's xa^{x-1}'.

No. No, it's not. That's the derivative with respect to a. You have to be cleverer than that – the only thing of that form (where the x is a power rather than a base) that you currently know how to differentiate is e^x – which is actually a good start!

Differentiating a^x

The trick to differentiating a^x is to turn it into something with e in the base – which you can do by saying $a = e^{\ln(a)}$. That means $a^x = \left(e^{\ln(a)}\right)^x = e^{x\ln(a)}$.

That's something you can differentiate with the chain rule: you can write $y = e^{x\ln(a)}$ as $y = e^u$, where $u = x\ln(a)$. Differentiating everything with respect to the right variable gives you $\dfrac{dy}{du} = e^u$ and $\dfrac{du}{dx} = \ln(a)$, so $\dfrac{dy}{dx} = e^u \ln(a) = e^{x\ln(a)} \ln(a) = a^x \ln(a)$.

Differentiating x^x

If differentiating a^x is tricky, x^x is worse still. The method is pretty much the same, though. Here's the recipe:

1. **Rewrite the expression with an e in the base.**

   You can write the x in the base as $e^{\ln(x)}$, so

   $$x^x = e^{x\ln(x)}$$

2. **Set up the chain rule.**

   Starting from $y = e^{x\ln(x)}$, take $y = e^u$ and $u = x\ln(x)$. You're going to find $\dfrac{dy}{dx}$, which is $\dfrac{dy}{du} \times \dfrac{du}{dx}$.

3. **Differentiate.**

   $\dfrac{dy}{du} = e^u$ and $\dfrac{du}{dx} = \ln(x) + 1$ (using the product rule).

4. **Multiply together and simplify.**

   $$\begin{aligned} \frac{dy}{dx} &= e^u\left(\ln(x)+1\right) \\ &= e^{x\ln(x)}\left(\ln(x)+1\right) \\ &= x^x\left(\ln(x)+1\right) \end{aligned}$$

Integrating by Substitution

The silly joke for integration by substitution goes like this: 'Imagine you're going through a horrible break-up: you don't want to see your *x* any more. It's all about *u*.' Don't look at me like that! It's a really handy way to remember what's going on.

Substitution has a number of forms, one of which is sometimes called the *reverse chain rule*, but I don't find splitting it up into different types especially helpful: there's one method that works for all types!

Substituting: The basic idea

The idea of integration by substitution is to change the variable you're working with, with the aim of turning something extremely nasty into something that's much, much simpler.

I'll start with a simple example, just to get the idea down. Say you need to work out $\int 3x^2 \cos\left(x^3\right) dx$. Here's what you do:

1. **If you're not told which substitution to use, find something ugly to set as *u*.**

   Here, the ugly thing is the x^3 in the brackets, so $u = x^3$.

2. **You want to convert from an *x*-integral to a *u*-integral. This means you need to work out what *du* is.**

   The chain rule says $du = \dfrac{du}{dx} dx$. $\dfrac{du}{dx}$ is easy to find (it's $3x^2$), so you have $du = 3x^2\ dx$.

3. **Substitute into the integral.**

   You have $3x^2\ dx$ in the integral, so you can replace that with du. You get $\int \cos\left(u\right) du$.

4. **If needed, use your substitution to get rid of any other *x*s hanging around.**

   Here, you don't.

5. **Integrate.**

   You get $\sin\left(u\right) + C$.

6. **Replace *u*s with *x*s again.**

   Your answer is $\sin\left(x^3\right) + C$.

If you were doing Further Maths, you'd really need to get good at spotting possible substitutions – but luckily, in Core 4, substitutions are usually either fairly obvious (there's one ugly thing that is very likely to be u) or you're told exactly what to do.

Working out what is ugly is a matter of taste and experience – good places to look for ugly things are in the bottoms of fractions and in the brackets after a function.

Changing limits

If you have limits to deal with when integrating by substitution, you have two options. The first is to work everything out as in the preceding section, get an answer in terms of x and do the old substitute-in-your-limits dance. It works perfectly fine, but I'd say it's a bit more work than you need to do. Slightly better is to turn your limits from xs into us and substitute in a little earlier. Here's an example:

$$\int_{-\frac{1}{2}\pi}^{\frac{1}{2}\pi} e^{\sin(x)+2} \cos(x)\mathrm{d}x$$

How much lovelier could it get? The answer is *none, none more lovely.* Here's what you do:

1. **Find the ugly thing.**

   Here, it's the power of e, so let $u = \sin(x)+2$. (With a little practice, you start to spot that the derivative of u is hanging around like a bad smell and you'll be able to get rid of both of the nasty bits in one go.)

2. **Change the integration so it's *du* rather than *dx*.**

   $du = \dfrac{du}{dx}\,dx$, and since $\dfrac{du}{dx} = \cos(x)$, $du = \cos(x)\,dx$.

3. **Put your *du* where it belongs.**

   $$\int_{x=-\frac{1}{2}\pi}^{x=\frac{1}{2}\pi} e^{u}du$$

   Notice that I've put '$x =$' with the limits – that's to remind me I need to replace them later (and because the du at the end implies that the limits are us).

4. **Replace the *x*-limits with *u*-limits.**

   If $x = \dfrac{1}{2}\pi, u = 3$; if $x = -\dfrac{1}{2}\pi, u = 1$.

5. Now integrate.

$$\int_1^3 e^u \, du = \left[e^u \right]_1^3 = e^3 - e = e(e-1)(e+1)$$

Told you it was lovely!

How about trying a nastier example? Use the substitution $x = 3\cos(u)$ to find the exact value of $\int_{\frac{3}{2}}^{\frac{3}{2}\sqrt{3}} \frac{1}{x^2 \sqrt{9-x^2}} \, dx$.

A thing to notice here: the substitution is 'the wrong way round' – most of the substitutions you can expect to see will be in the form $u = f(x)$; this one has $x = f(u)$. It isn't all that much different, though – you still need to get rid of your x. (Incidentally, you can expect to be given the right substitution to use for anything this nasty.) Here's what to do:

1. **Replace the *dx* with something involving *du*, which you get by differentiating what you're about to substitute in.**

   As usual, $du = \frac{du}{dx} \, dx$. You can get $\frac{dx}{du} = -3\sin(u)$ by differentiating and then use the fact that $\frac{du}{dx} = \frac{1}{\left(\frac{dx}{du} \right)}$ to say $du = \frac{1}{-3\sin(u)} \, dx$.

2. **Substitute for *x* in the integrand using your substitution.**

   Here, the integrand becomes $\frac{1}{9\cos^2(u)\sqrt{9-9\cos^2(u)}}$, which simplifies to $\frac{1}{27\cos^2(u)\sin(u)}$.

3. **Put your integral together and simplify further if possible.**

   Ignoring the limits for the moment, you want to replace $\frac{1}{-3\sin(u)} \, dx$ with *du*, which you can do: $\int \frac{1}{27\cos^2(u)\sin(u)} \, dx = -\int \frac{1}{9\cos^2(u)} \, du$, or, better yet, $-\frac{1}{9}\int \sec^2(u) \, du$.

4. **Change your limits, again using your substitution.**

   With this one, when $x = \frac{3}{2}$, $u = \frac{1}{3}\pi$, and when $x = \frac{3}{2}\sqrt{3}$, $u = \frac{1}{6}\pi$. Make sure you put them in the right places – the u in each slot needs to correspond to the x that was originally there!

5. Now do the integration!

$$-\frac{1}{9}\int_{\frac{1}{3}\pi}^{\frac{1}{6}\pi}\sec^2(u)\,du = -\frac{1}{9}\Big[\tan(u)\Big]_{\frac{1}{3}\pi}^{\frac{1}{6}\pi}$$

$$= -\frac{1}{9}\left(\frac{1}{3}\sqrt{3} - \sqrt{3}\right)$$

$$= \frac{2}{27}\sqrt{3}$$

If your integral doesn't have limits, you'll need to convert your *us* back into *xs* using the original substitution.

Integrating by Parts

The integration by parts formula is a close cousin of the product rule for differentiation (in fact, if you integrate the product rule, you get the parts formula). You use integration by parts in similar circumstances, when you have two things multiplied together.

In this section, I show you the rule, which involves differentiating one of the factors and integrating the other, and I show you how to decide which factor is which. I give you a run-of-the-mill example and then a few less obvious ones.

Knowing the rule

You don't really *need* to know the parts formula – it's in the formula book – but it comes up often enough that you might as well learn it. (If you have your eyes on a university course with a maths component, chances are you'll need to know it, so learning it now may make for a slightly less stressful future.) What you're given is something like this:

$$\int u\frac{dv}{dx}\,dx = uv - \int v\frac{du}{dx}\,dx$$

(If you rearrange that and differentiate, you get $u\dfrac{dv}{dx} + v\dfrac{du}{dx} = \dfrac{d}{dx}(uv)$, which ought to be familiar to you as the product rule. Because integration and differentiation are inverses, it shouldn't be a huge surprise that integrating the product rule gives you something that works.)

A parts question will involve integrating something with two factors, one of which $\left(\dfrac{dv}{dx}\right)$ you'll choose to integrate to get v, and one of which (u) you'll differentiate to get $\dfrac{du}{dx}$; after you do that, you just need to replace the letters as appropriate, do the final integral and simplify your answer.

(Hahaha! 'Just'. Of course, that's where the difficulty usually comes in.)

Deciding which bit is which

There's a fairly simple rule for deciding which bit to integrate and which bit to differentiate: you differentiate whichever bit will get simpler when you differentiate it! More concretely:

> ✔ **If there's a logarithm:** Assuming you're using parts, you're almost certain to differentiate $\ln(x)$ rather than integrate. The natural log is surprisingly tricky to integrate (but see later in this chapter for how)!

> ✔ **Otherwise, if there's a power of *x*:** The u will be x to the power of whatever it is.

> ✔ **Otherwise:** At A level, chances are you can do it without parts (but if you do it with parts, which is which probably doesn't matter).

Parts in action

A garden-variety parts question may ask you to find something like $\int 4\theta \cos(2\theta)\,d\theta$. Here's how to approach it (Figure 17-4 will help):

1. **Decide what to integrate and what to differentiate.**

   Here, the 4θ will become simpler if you differentiate it, while $\cos(2\theta)$ doesn't get *much* trickier through integration, so set $u = 4\theta$ and $\frac{dv}{d\theta} = \cos(2\theta)$.

2. **Integrate and differentiate as needed.**

   $\frac{du}{d\theta} = 4$ and $v = \frac{1}{2}\sin(2\theta)$.

3. **Substitute your values into the parts formula.**

   $$uv - \int v\frac{du}{d\theta}\,d\theta = (4\theta)\left(\frac{1}{2}\sin(2\theta)\right) - \int\left(\frac{1}{2}\sin(2\theta)\times 4\right)d\theta$$

4. **Simplify and do the integration.**

   The first term is $2\theta\sin(2\theta)$; the integral is $\int 2\sin(2\theta)\,d\theta$, which integrates to $-\cos(2\theta)$.

5. **Write it all out!**

   You get $2\theta\sin(2\theta) + \cos(2\theta) + C$.

Figure 17-4:
Worked
example
of parts
question.

$$u = 4\theta \qquad \frac{dv}{d\theta} = \cos(2\theta)$$

$$\frac{du}{d\theta} = 4 \qquad v = \frac{1}{2}\sin(2\theta)$$

$$2\theta \sin(2\theta) - \int 2\sin(2\theta)\,d\theta$$
$$= 2\theta \sin(2\theta) + \cos(2\theta) + C$$

© John Wiley & Sons, Inc.

Sneaky tricks

Applying integration by parts is usually as straightforward as sticking every-thing in the formula, cranking the handle and writing down the answer, which is lovely in terms of getting marks but hardly very interesting! Every so often, a question that needs a bit more thought crops up – for example, something where you need to apply the parts formula repeatedly (very common) or to integrate something with a natural logarithm in (less common, hence a blood-bath when it *does* come up).

Repeated parts

A typical repeated-parts question would ask you to integrate something like $x^2 e^{-3x}$ (with respect to x). Here's how you go about it:

1. **Spot that it's two things multiplied together. Pick something that'll get simpler as u and pick the other thing as $\frac{dv}{dx}$.**

   Here, $u = x^2$ and $\frac{dv}{dx} = e^{-3x}$.

2. **Differentiate and integrate as normal.**

   $\frac{du}{dx} = 2x$ and $v = -\frac{1}{3}e^{-3x}$.

3. **Apply the parts formula.**

   $$\int x^2 e^{-3x}\,dx = -\frac{1}{3}x^2 e^{-3x} + \frac{2}{3}\int x e^{-3x}\,dx$$

4. **You still have to integrate two things multiplied together! Work out the second integral using parts again.**

   $U = x$ and $\frac{dV}{dx} = e^{-3x}$, so $\frac{dU}{dx} = 1$ and $V = -\frac{1}{3}e^{-3x}$. The second integral is

   $\left(-\frac{1}{3}x e^{-3x} + \frac{1}{3}\int e^{-3x}\,dx \right)$.

5. **Finish off the second integral!**

   That's $\left(-\frac{1}{3}x e^{-3x} - \frac{1}{9}e^{-3x} \right)$.

6. Substitute that back into your result from Step 3.

$$\int x^2 e^{-3x}\, dx = -\frac{1}{3}x^2 e^{-3x} + \frac{2}{3}\left(-\frac{1}{3}xe^{-3x} - \frac{1}{9}e^{-3x}\right) + C$$

7. Tidy that monster up!

I'd take $-\frac{1}{27}e^{-3x}$ out to leave me with $-\frac{1}{27}e^{-3x}\left(9x^2 + 6x + 2\right) + C$.

It's *very* easy to lose a minus sign in this sort of question. All those times your teacher nagged you about laying your work out neatly? This is where it pays off – the easier things are to read, the easier it is to spot the errors you're bound to make. (I mention this because *I* lost a minus sign in working that out, but I found it by careful checking.)

Integrating a^x

If you know how to differentiate a^x (having learnt it in the earlier section 'Differentiating a^x' or elsewhere), you know the trick you need to integrate it, too! Here's how it goes:

1. Rewrite a^x as $e^{x\ln(a)}$.

2. This includes a linear expression (because $\ln(a)$ is a constant), so you can integrate it easily: $\int e^{x\ln(a)}\, dx = \frac{1}{\ln(a)}e^{x\ln(a)} + C$.

3. Tidy up by turning the e bit back into something nicer: $\frac{1}{\ln(a)}a^x + C$.

Integrating $\ln(x)$

Integrating a natural log is the kind of question that looks simple until you try it. At that point, you remember you can differentiate it, but you've (probably) no idea how to integrate it. Luckily, that's also the clue about how to do it – you can integrate $1 \times \ln(x)$ using the parts formula! (In fact, you use the same trick for any power of x multiplied by a logarithm.) Here's how:

1. Let u be the logarithm, and let $\frac{dv}{dx}$ be the power of x.

To work out $\int \ln(x) \times 1 dx$, set u to be the logarithm, giving you $u = \ln(x)$, and set $\frac{dv}{dx}$ to be the power of x, giving you $\frac{dv}{dx} = x^0$.

2. Differentiate u to get $\frac{du}{dx}$, and integrate $\frac{dv}{dx}$ to get v.

Here, $\frac{du}{dx} = \frac{1}{x}$ and $v = x$.

3. Apply the parts formula.

The integral is $uv - \int v \frac{du}{dx} dx = x\ln(x) - \int 1 dx$.

4. Integrate the last bit.

You get an answer of $x\ln(x) - x + C$.

If you wanted to tidy that up as $x\left(\ln\left(x\right)-1\right)+C$, you'd look like a mathematical rock star!

Similarly, to integrate $x^{2}\ln\left(x\right)$, you'd have $u=\ln\left(x\right)$ and $\dfrac{du}{dx}=\dfrac{1}{x}$ as before, while $\dfrac{dv}{dx}=x^{2}$ and $v=\dfrac{1}{3}x^{3}$. That gives you an integral of

$$\int x^{2}\ln\left(x\right)dx=\frac{1}{3}x^{3}\ln\left(x\right)-\frac{1}{9}x^{3}+C$$

$$=\frac{1}{9}x^{3}\left(3\ln\left(x\right)-1\right)+C$$

Integration by Trig Identity

In some Core 4 exams, there's an audible thud about half an hour in as all the candidates turn the page at once, they see the next question, and every heart in the room sinks. It's the one problem they'd hoped wouldn't come up: the trigonometric integral.

And I'm not talking about one of the $x\cos\left(x\right)$ cakewalks that parts will deal with in a trice. It's not even one of the $e^{x}\sin\left(e^{x}\right)$ monsters you can substitute in. But here is pure trig: something like $\int\cot^{2}\left(x\right)dx$ or $\int\sin\left(5x\right)\cos\left(3x\right)dx$.

You're going to need some identities, my friend. Luckily, the formula book is brimming with them.

Squares of tan and cot

Out of the squared trig functions, $\tan^{2}\left(x\right)$ and $\cot^{2}\left(x\right)$ are probably the least complicated to integrate, apart from $\sec^{2}\left(x\right)$ (which integrates directly to $\tan\left(x\right)+C$) and $\mathrm{cosec}^{2}\left(x\right)$ (which gives you $-\cot\left(x\right)+C$), both of which are in the formula book. In fact, those two are the key to integrating $\tan^{2}\left(x\right)$ and $\cot^{2}\left(x\right)$, because they're linked by the identities from Chapter 12: $\tan^{2}\left(x\right)\equiv\sec^{2}\left(x\right)-1$ and $\cot^{2}\left(x\right)\equiv\mathrm{cosec}^{2}\left(x\right)-1$. That makes integrating them barely worth a recipe:

1. **Replace your squared function with the other side of the identity.**

$$\int\tan^{2}\left(x\right)dx=\int\sec^{2}\left(x\right)-1dx$$

2. **Integrate.**

$$\int\sec^{2}\left(x\right)-1dx=\tan\left(x\right)-x+C$$

I told you it was (relatively) simple. Unfortunately, the other squared functions are a bit trickier.

Double angles

Integrating $\sin^2(x)$ or $\cos^2(x)$ is a nasty thing to ask anyone to do. There's a classical way to do it (in this section) and a smart alternative (in the next).

The classical way hinges on the identity $\cos(2x) \equiv \cos^2(x) - \sin^2(x)$, and it relies on noticing that $\cos(2x)$ is comparatively easy to integrate. If you can rewrite your squared trig function in terms of $\cos(2x)$, you're away! Here's the plan for integrating $\sin^2(x)$ (the method for $\cos^2(x)$ is almost identical):

1. **Write down the double-angle identity: $\cos(2x) \equiv \cos^2(x) - \sin^2(x)$.**

2. **Replace the squared thing you don't want using the identity $\sin^2(x) + \cos^2(x) \equiv 1$.**

   Here, you can say $\cos^2(x) \equiv 1 - \sin^2(x)$, so $\cos(2x) \equiv 1 - 2\sin^2(x)$.

3. **Rearrange to get your squared thing on its own.**

   $$\sin^2(x) \equiv \frac{1}{2}\big(1 - \cos(2x)\big)$$

4. **Integrate the right-hand side!**

   Here, you get

   $$\int \sin^2(x)\,dx = \frac{1}{2}\int 1 - \cos(2x)\,dx$$
   $$= \frac{1}{2}\left(x - \frac{1}{2}\sin(2x)\right) + C$$

There are several ways to tidy that up – my favourite is $\frac{1}{4}\big(2x - \sin(2x)\big) + C$, but as long as it works out to the same thing, the answer is correct.

Products of sine and cosine

If you do the Solomon papers as a means of revising – something I'd definitely recommend after you run out of ordinary past papers – you'll frequently see questions like 'What is $\int \sin(4x)\cos(2x)\,dx$?'

If you're *very* clever about it, you can do that problem by repeatedly applying the parts rule – but this isn't the place for doing things very cleverly. I like to do things the easy way, which is to look in the formula book.

Under 'Trigonometric identities' in the Core 3 section, after the compound angle formulas, you'll see four more identities you may not have covered in class (you can find more on them in Chapter 12 if you're interested):

$$\sin(A) + \sin(B) \equiv 2\sin\left(\frac{A+B}{2}\right)\cos\left(\frac{A-B}{2}\right)$$

$$\sin(A) - \sin(B) \equiv 2\cos\left(\frac{A+B}{2}\right)\sin\left(\frac{A-B}{2}\right)$$

$$\cos(A) + \cos(B) \equiv 2\cos\left(\frac{A+B}{2}\right)\cos\left(\frac{A-B}{2}\right)$$

$$\cos(A) - \cos(B) \equiv -2\sin\left(\frac{A+B}{2}\right)\sin\left(\frac{A-B}{2}\right)$$

How do these help? The key is on the right-hand side – if you can find A and B that conform to the question you're asked, you can turn the nasty multiplication into a simpler addition.

To integrate $\sin(4x)\cos(2x)$, you do this:

1. **Look for the right-hand side that's in the same kind of form as what you've got.**

   Here, it's the first identity – sine of something multiplied by cosine of something.

2. **Match up your As and Bs with the xs in brackets by setting up simultaneous equations.**

   You're trying to write $\sin(4x)\cos(2x)$ in the form $\sin\left(\frac{A+B}{2}\right)\cos\left(\frac{A-B}{2}\right)$. The argument of the first factor suggests $\frac{A+B}{2} = 4x$; the second factor gives you $\frac{A-B}{2} = 2x$.

3. **Solve the simultaneous equations.**

   $A = 6x$ and $B = 2x$.

4. **Figure out how to get your original integrand on the right-hand side.**

   You know that $\sin(6x) + \sin(2x) \equiv 2\sin(4x)\cos(2x)$, which you'll need to halve to get your original integrand on the right.

5. **Integrate the freshly halved left-hand side!**

$$\frac{1}{2}\int \sin(6x) + \sin(2x)\,dx = -\frac{1}{12}\cos(6x) - \frac{1}{4}\cos(2x) + C$$

TIP

If you had picked the second identity, treating it as a cosine multiplied by a sine, you would still have gotten the same answer in the end, at the cost of dealing with a few extra minus signs.

I mentioned earlier that you could use this trick to integrate things like $\cos^2(x)$ – here's a recipe for that, too:

1. **Look for the right-hand side with the appropriate functions multiplied together.**

   Here, you want $2\cos\left(\dfrac{A+B}{2}\right)\cos\left(\dfrac{A-B}{2}\right)$.

2. **Work out A and B that give you what you need in the brackets.**

   If $\dfrac{A+B}{2} = x$ and $\dfrac{A-B}{2} = x$, then $A = 2x$ and $B = 0$.

3. **Put these values into the left-hand side of the identity.**

   $\cos(A) + \cos(B) \equiv 2\cos\left(\dfrac{A+B}{2}\right)\cos\left(\dfrac{A-B}{2}\right)$, so the left-hand side is

   $\cos(2x) + \cos(0)$

4. **Tidy that up to get an expression for what you started with.**

   $\cos^2(x) = \dfrac{1}{2}\left(\cos(2x) + 1\right)$

5. **Integrate this!**

   You get $\dfrac{1}{4}\sin(2x) + \dfrac{1}{2}x + C$, or any number of equivalents.

Deciding Which Integration Rule to Use

Often, the hardest bit of an integration question is deciding which rule is the right one to use. There's no one-size-fits-all recipe that I'm aware of, but I can offer some general hints:

- ✔ If the question tells you which method to use, try using that method. Call me old-fashioned if you like, but I reckon taking advantage of hints is a good idea.

- ✔ Similarly, if the question says 'hence', or 'hence or otherwise', there's something in the previous part you should use. (Often, you'll have differentiated something to get the thing you're about to integrate, so integrating back would take you to whatever you differentiated to start with!)

- ✔ If you've got something awful on the bottom of a fraction or in a bracket, try substituting u for the awful thing.

✔ If you've got two seemingly unrelated things multiplied together, especially with an x or a log, try integration by parts.

✔ If you've got a squared trig function that isn't in the formula book, try using identities to turn it into one that is ($\sec^2(x)$ or $\operatorname{cosec}^2(x)$) or something you know ($\cos(2x)$).

✔ If you've got two trig functions multiplied together, use the formulas in your book to split them into an addition problem.

Chapter 18

Overcoming Evil Questions

· ·

· ·

*E*vil questions? I don't think things like differential equations are necessarily *evil* questions . . . they're just misunderstood. Often badly misunderstood, if I'm honest. All the same, when you get to Core 4, there are a few topics that don't really feel as though they're linked to what you've done before. (In fact, they are – it's just that the links don't become clear until you're a few branches further up the maths tree.)

In any case, these so-called evil questions can really get in the way of your grade, so in this chapter, I show you how to deal with them. You see a few new ways of defining curves – implicitly (which means 'the *x*s and *y*s are all mixed together') and parametrically (which means you define *x* and *y* separately, in terms of another variable). You also take integration into the third dimension with volumes of revolution, and you start exploring differential equations – something that makes people like me nod and say, 'That's *proper* maths.' (Not in a condescending way, you understand – but if you go on to do a maths degree, you'll spend a lot of time getting to the bottom of differential equations, which help tie together a lot of what you've been doing up until now.)

Parametric Curves

A *parametric* curve is one where the *x* and *y* coordinates are specified separately, in terms of some other variable – a *parameter*, hence the name. The parameter is usually *t* or occasionally θ, but that's just convention – there's no reason the parameter can't be another letter if you want.

This approach means you can define complicated curves much more easily. For example, one of the EdExcel past papers (January 2007) defined a curve as $x = 7\cos(t) - \cos(7t)$ and $y = 7\sin(t) - \sin(7t)$. If you plot that, you get the curve in Figure 18-1: a lovely cloud! For the sake of accuracy, the question defined the curve only for $\frac{1}{8}\pi < t < \frac{1}{3}\pi$, so you were looking at only a bit of one of the bulges – but I prefer the whole graph!

Parametric curves allow you to break free of the 'one y-value per x-value' constraint that functions obey; by expressing x and y as functions of a parameter (giving one x-value and one y-value for each t-value), you open up a whole new world of possibilities. Any reasonable curve can be expressed parametrically, from a circle to the Bat-Signal and beyond.

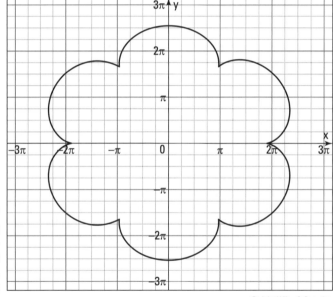

Figure 18-1:
A para-
metric
curve.

© John Wiley & Sons, Inc.

In this section, I show you how this curve is made and how to find the gradient at (almost) any point.

I say *almost* any point because the gradient isn't defined at several points. At the sharp bits, the maths breaks down, so I'm going to ignore those. You can ask your university lecturers about them later on.

Finding points

To find a point on a parametrically defined curve, you put a value of t into both equations and end up with a value for x and a value for y. For example, if $t = \frac{1}{4}\pi$, then $x = 7\cos\left(\frac{1}{4}\pi\right) - \cos\left(\frac{7}{4}\pi\right)$, which is $3\sqrt{2}$. Meanwhile, $y = 7\sin\left(\frac{1}{4}\pi\right) - \sin\left(\frac{7}{4}\pi\right) = 4\sqrt{2}$. The point $\left(3\sqrt{2}, 4\sqrt{2}\right)$ corresponds to the t-value of $\frac{1}{4}\pi$. This is a typical introduction to a parametric curve equation, and it's quite a nice way to pick up a mark or two.

Finding points that satisfy conditions on x or y is a bit more involved (and not really feasible with the cloud example) but not *much* more involved: you can treat the condition as a third equation and solve the equations simultaneously.

A typical example might be set up as $x = \ln(2t)$ and $y = t^2 - 8$ for $t > 0$. To find where this crosses the x-axis, you would solve for where $y = 0$. If $0 = t^2 - 8$, then $t^2 = 8$ and $t = 2\sqrt{2}$ (note that the negative square root isn't in the domain). Putting the t-value into the x equation gives you $x = \ln(4\sqrt{2})$ or $x = \frac{5}{2}\ln(2)$. This curve crosses the x-axis at $\left(\frac{5}{2}\ln(2), 0\right)$.

Finding gradients

The key to finding the gradient – the change in y as x varies – of a parametric curve is to remember that $\frac{dy}{dx} = \frac{dy}{dt} \times \frac{dt}{dx}$. That's the chain rule again, with the dts lining up nicely on the diagonal! If you also know that $\frac{dt}{dx} = \frac{1}{\left(\frac{dx}{dt}\right)}$, you can rewrite it as $\frac{dy}{dx} = \frac{dy}{dt} \div \frac{dx}{dt}$. (While it *looks* like the dts 'cancel out', that's not really what's happening: derivatives aren't fractions, and you have to go through the chain rule bureaucracy to be sure of getting things right.)

If you want to find the gradient of the cloud curve defined as $x = 7\cos(t) - \cos(7t)$ and $y = 7\sin(t) - \sin(7t)$ (from the previous section), go through the following recipe:

1. **Differentiate the y equation with respect to t.**

   You get $\frac{dy}{dt} = 7\cos(t) - 7\cos(7t)$.

2. **Differentiate the x equation with respect to t.**

   You get $\frac{dx}{dt} = -7\sin(t) + 7\sin(7t)$.

3. Divide the *y* one by the *x* one.

You get $\dfrac{dy}{dx} = \dfrac{7\cos(t) - 7\cos(7t)}{-7\sin(t) + 7\sin(7t)}$, with a factor of 7 you can take

out to get $\dfrac{dy}{dx} = \dfrac{\cos(t) - \cos(7t)}{-\sin(t) + \sin(7t)}$.

4. If you want the gradient in terms of *t*, you're done. If, instead, you were given a specific value of *t*, you'd substitute it in.

At $t = \dfrac{1}{6}\pi$, for example, the gradient is $-\sqrt{3}$. See Figure 18-2.

When you differentiate parametrically to get $\dfrac{dy}{dx}$, you end up with an expression for how *y* varies as you change *x* – all written in terms of the parameter *t*. You can, in many cases, rewrite everything in terms of *x*, but I really can't think of a good reason to. When you're writing the equation of a tangent or normal, however, you need everything to be in terms of *x* and *y*.

If needed, you can use this technique to work out tangents, normals and stationary points.

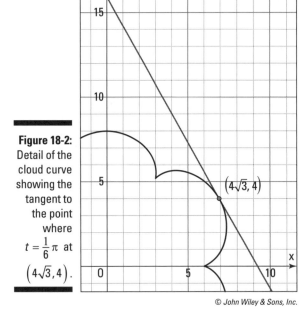

Figure 18-2: Detail of the cloud curve showing the tangent to the point where $t = \dfrac{1}{6}\pi$ at $\left(4\sqrt{3}, 4\right)$.

You've already done some of the work to find the tangent to the cloud curve at the point where $t = \frac{1}{6}\pi$, in that you know the gradient there. All you need are the x- and y-coordinates, which you can put in the straight line equation and *bingo!* You're done. A recipe for this:

1. **Find the gradient of the curve $\left(\dfrac{dy}{dx}\right)$ at the given point.**

   Using the recipe from earlier in this section, the gradient is $-\sqrt{3}$.

2. **If you need the normal, find the perpendicular gradient.**

   You don't need it here, but if you did, you'd find the negative reciprocal; it's $\dfrac{1}{\sqrt{3}}$.

3. **Find the required point (x_0, y_0) on the curve by substituting t into the x and y equations.**

   Because $x = 7\cos(t) - \cos(7t)$, when $t = \frac{1}{6}\pi$, $x_0 = 4\sqrt{3}$; because $y = 7\sin(t) - \sin(7t)$, when $t = \frac{1}{6}\pi$, $y_0 = 4$.

4. **Think back to the halcyon days of Core 1 and the straight line equation, and put the correct values in it.**

   You have $m = -\sqrt{3}$, $x_0 = 4\sqrt{3}$ and $y_0 = 4$, so your straight line equation is $y - 4 = -\sqrt{3}\left(x - 4\sqrt{3}\right)$. You can rearrange that to be $\sqrt{3}x + y - 16 = 0$.

Converting to Cartesian form

It's probably never occurred to you that the normal way of writing a curve, with just ys and xs in it, has a name, but it does. It's called the *Cartesian* form, after René Descartes. (Yes, *he thought; therefore, he was* a philosopher – but in those days, you could follow several careers. He was also at the forefront of seventeenth-century French physics, and he invented the coordinate system based on x, y and z in his spare time.) The story goes that equations with just x and y in them were especially pleasing to Descartes because he was from France and thus didn't like t.

Finding the Cartesian form of a parametrically defined curve is sometimes a case of eliminating t in a simple and straightforward way; other times, it involves a lot of head-scratching and possibly a hammer. In this section, I show you one of each.

First, a relatively simple one (see Figure 18-3): $x = t^3 + 1$ and $y = \dfrac{4}{t}$ for $t \neq 0$, and you want the Cartesian equation of the curve in the form $y = f(x)$. Here's what I'd do:

1. **Get t on its own in the simpler equation.**

   $$t = \frac{4}{y}$$

2. Substitute into the other equation.

$$x = \left(\frac{4}{y}\right)^3 + 1$$

$$x = \frac{64}{y^3} + 1$$

3. Rearrange into the form they asked for.

$$x - 1 = \frac{64}{y^3}, \text{ so } y^3 = \frac{64}{x-1} \text{ and } y = \sqrt[3]{\frac{64}{x-1}}.$$

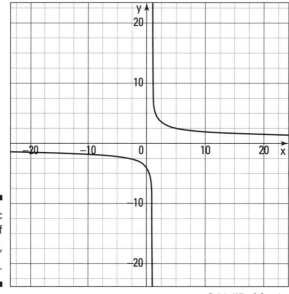

Figure 18-3:
The graph of
$x = t^3 + 1$,
$y = \frac{4}{t}$.

© John Wiley & Sons, Inc.

WARNING!

If they ask for the Cartesian equation in the form $y = f(x)$, it's a good idea to check that what comes out of your working is actually a function. In this case, there's no problem, but you should be especially careful around square roots; make sure that you pick correctly between positive and negative by checking against a *t*-value.

This method is lovely, as long as you don't have nasty functions involved – you don't want to be playing about with sines and inverse cosines and so on in this kind of question – much better to use some trig identities and make life simpler for yourself!

For example, if you have $x = \tan^2(t)$ and $y = \sin(t)$ for $0 < t < \frac{1}{2}\pi$ (see Figure 18-4) and you need to find a Cartesian equation of the curve in the form $y^2 = f(x)$ (as you would have done in the June 2006 EdExcel Core 4 paper), here's what I'd suggest:

1. **Scour the question for clues. Here, y^2 appears significant, so that's a good place to start.**

$$y^2 = \sin^2(t)$$

2. **Work out how that can relate to the other equation.**

$\tan^2(t) \equiv \dfrac{\sin^2(t)}{\cos^2(t)}$, which looks promising, especially when you remember the bottom is $1 - \sin^2(t)$.

3. **Substitute and rearrange as necessary.**

$x = \dfrac{y^2}{1 - y^2}$, so $x - xy^2 = y^2$, and $y^2(1 + x) = x$. That means $y^2 = \dfrac{x}{1 + x}$.

WARNING!

This is what I was talking about in the last Warning! The naïve answer of $y = \pm\sqrt{\dfrac{x}{1 + x}}$ has a multi-valued right-hand side, which means y isn't a function of x. This is why they've asked for y^2; this works just fine.

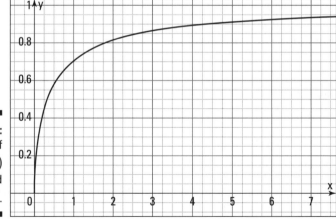

Figure 18-4:
The graph of $x = \tan^2(t)$ and $y = \sin(t)$.

© John Wiley & Sons, Inc.

There are dozens of tricks you may need to employ in this sort of question, and it's impossible to cover them all. Strategies like 'spot the difference', 'what do these things have in common?' and 'how can I make this less ugly?' are all good things to think about. While exam boards are, by their nature, full of big blue meanies, they *are* constrained by the syllabus and Ofqual, so they're not meant to ask you anything it's unreasonable for you to figure out.

Integrating parametrically

Being asked to integrate a parametrically defined function is unusual, but it does happen. Again, it's the kind of thing that might be thrown in to keep you on your toes. Luckily, if you can integrate by substitution, you can integrate parametric functions just nicely.

It helps to have a plan going into this kind of question. You're used to seeing y as a function of x and merrily integrating with respect to x. With a parametric curve, though, you integrate with respect to t, the link between y and x. It screams out 'substitution', doesn't it?

For example, say a curve is defined as $x = \tan(t)$ and $y = \sin(t)$ for $0 < t < \frac{1}{4}\pi$. Here's how you'd find the area beneath it:

1. **Change from *dx* to *dt* by differentiating your *x* equation.**

   The chain rule says $dx = \dfrac{dx}{dt}\,dt$, and differentiating says $\dfrac{dx}{dt} = \sec^2(t)$, so $dx = \sec^2(t)\,dt$.

2. **Replace *y* and *dx* in the integral with their equivalent expressions in *t*, and put limits in terms of *t*.**

   The area you wanted originally (as in Figure 18-5) was $\int_0^1 y\,dx$. Changing to t, you get $\int_0^{\frac{1}{4}\pi} \sin(t)\sec^2(t)\,dt$.

3. **Integrate.**

   For this one, the substitution $u = \cos(t)$ turns the integral into $-\int_1^{\frac{1}{2}\sqrt{2}} u^{-2}\,du$, which works out to be $\left[u^{-1}\right]_1^{\frac{1}{2}\sqrt{2}}$, or $\sqrt{2} - 1$.

For this sort of question, unless the integral is quite straightforward, you can expect a bit of guidance. I don't think you'd usually be expected to find that cosine substitution on your own.

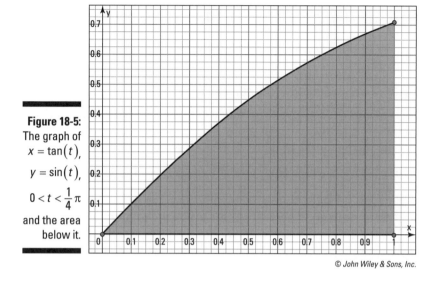

Figure 18-5:
The graph of
$x = \tan(t)$,

$y = \sin(t)$,

$0 < t < \frac{1}{4}\pi$

and the area
below it.

Implicit Curves

A curve is defined *explicitly* if it's in the form $y = f(x)$ or (rarely) $x = f(y)$ – that is to say, one variable is defined completely in terms of the other. Almost every curve you've seen so far has been like this, with one exception: the circle.

An equation of a circle is $(x-a)^2 + (y-b)^2 = r^2$, which is *implicitly* defined – neither variable is on its own; they're mixed up together. (In some implicit curves, you have xs and ys on both sides of the equation.)

You could, with a bit of work, figure out an explicit equation for a circle. (Something like $y = b \pm \sqrt{r^2 - (x-a)^2}$ would do the trick – it's not especially pretty, is it? It also loses the Pythagorean clarity of the classical version.) But in many cases, finding an explicit equation is not possible.

A typical implicit-curve question may define a curve as $x^3 - 2y^2 = 3xy$ and ask you to find the points where $x = 5$ as well as the gradient at those points. This section shows you how!

Finding points on an implicit curve

To find a point or points on an implicitly defined curve, follow these simple steps:

1. **Substitute what you know into the equation.**

   The given curve is $x^3 - 2y^2 = 3xy$ and you know $x = 5$, so you can replace all the xs with 5 to get $125 - 2y^2 = 15y$.

2. **Solve the resulting equation.**

   $2y^2 + 15y - 125 = 0$, which factorises as $(y - 5)(2y + 25)$, so $y = 5$ or $y = -\dfrac{25}{2}$.

3. **If necessary, substitute back into your original equation(s) to get the other coordinate.**

   Here, you know $x = 5$. Your answers are $(5, 5)$ and $\left(5, -\dfrac{25}{2}\right)$.

Differentiating implicitly

Differentiating an implicit curve – to find its stationary points or its gradient – can be messy. You've probably not given much thought to what happens on the left-hand side when you differentiate, say, $y = x^3$ to get $\dfrac{dy}{dx} = 3x^2$; it's quite reasonable to focus on the right. However, the left-hand side is important: when you differentiate a y with respect to x, you get $\dfrac{dy}{dx}$. All you need to do is make sure you differentiate everything, including the ys, and then rearrange to get a nice answer!

Be prepared to have xs, ys and $\dfrac{dy}{dx}$s all mixed together as you work through an implicit differentiation problem. You can sort everything out in the end.

Here's how to differentiate $x^3 - 2y^2 = 3xy$, which is the example from earlier, by taking the derivative with respect to x of both sides:

✔ The first term on the left-hand side doesn't have a y in it, so you differentiate it as usual. You get $3x^2$.

✔ The second term has only ys in it. If you call it $z = -2y^2$, then the chain rule says $\dfrac{dz}{dx} = \dfrac{dz}{dy} \times \dfrac{dy}{dx}$, which is $-4y\dfrac{dy}{dx}$.

✔ The right-hand side requires the product rule, which says $\dfrac{d}{dx}(uv) = u\dfrac{dv}{dx} + v\dfrac{du}{dx}$. If you pick $u = 3x$ and $v = y$, then $\dfrac{du}{dx} = 3$ and $\dfrac{dv}{dx} = \dfrac{dy}{dx}$. The derivative is $(3x)\left(\dfrac{dy}{dx}\right) + (y)(3) = 3x\dfrac{dy}{dx} + 3y$.

Putting it all together gives you $3x^2 - 4y\dfrac{dy}{dx} = 3x\dfrac{dy}{dx} + 3y$. If you need to get $\dfrac{dy}{dx}$ on its own, you can treat that as a variable and shuffle it around. You may go to $3x^2 - 3y = 3x\dfrac{dy}{dx} + 4y\dfrac{dy}{dx}$, then factorise the right-hand side: $3x^2 - 3y = \dfrac{dy}{dx}(3x + 4y)$. Dividing by $(3x + 4y)$ gives you the final result, $\dfrac{dy}{dx} = \dfrac{3x^2 - 3y}{3x + 4y}$.

In implicit differentiation, it's *totally normal* to get a derivative in terms of x and y, all mixed up together. Don't panic!

After you know that equation, you can find the derivatives at $(5, 5)$ and $\left(5, -\dfrac{25}{2}\right)$ by substituting those values into what you worked out. At $(5, 5)$, you get $\dfrac{dy}{dx} = \dfrac{12}{7}$; at $\left(5, -\dfrac{25}{2}\right)$, you get $\dfrac{dy}{dx} = -\dfrac{45}{14}$.

When you're given an implicit curve and asked to find a tangent or normal, it's easy to get lost in which letter means what – but it's actually less awful than you'd expect. In fact, it's not *all* that different to what you've been doing all along: you find the gradient you need, then use the point you're given to find the equation of a line.

Whenever you're trying to find a tangent line – to parametric, implicit or common-or-garden curves, you need to find $\dfrac{dy}{dx}$; the lines you want are always in the simple 2D system of xs and ys that you've been playing with forever.

For example, if you're given the curve $x^2 + 3xy - 2y^3 = 8$ and asked find the normal at $\left(-2\sqrt{2}, 0\right)$, here's what you do:

1. **Differentiate implicitly.**

   The first term on the left simply becomes $2x$; the second, using the product rule, becomes $3y + 3x\dfrac{dy}{dx}$; the third, using the chain rule, is $-6y^2\dfrac{dy}{dx}$, and the right-hand side (being a constant) vanishes. You get

   $$2x + 3y + 3x\dfrac{dy}{dx} - 6y^2\dfrac{dy}{dx} = 0$$

2. **Find the value of $\dfrac{dy}{dx}$ at the point you're interested in to get the gradient of the tangent.**

   You don't really need to rearrange to get $\dfrac{dy}{dx}$ on its own unless you're explicitly asked for it – you can substitute x and y in directly and then solve if you prefer. If you do, you get $-4\sqrt{2} - 6\sqrt{2}\dfrac{dy}{dx} = 0$, or $\dfrac{dy}{dx} = -\dfrac{2}{3}$.

3. **Decide whether you want the gradient or the normal. If you need the normal, take the negative reciprocal of the result from Step 2.**

Here, you want the normal, which is $\frac{3}{2}$.

4. **Use your gradient and the point to work out the equation of the line.**

$(y-0) = \frac{3}{2}(x+2\sqrt{2})$, so $2y = 3x + 6\sqrt{2}$, or $3x - 2y + 6\sqrt{2} = 0$. Other equations are possible, unless the question specifies the form.

A question may also ask where a curve has a given gradient. That's a little bit trickier. Say you still have $x^2 + 3xy - 2y^3 = 8$, but you want to know the point where the curve is parallel to the y-axis in the first quadrant, which is a fancy way of saying $\frac{dx}{dy} = 0$ and x and y are both positive. (Parallel to the x-axis would be $\frac{dy}{dx} = 0$.)

1. **Differentiate implicitly.**

$$2x + 3y + 3x\frac{dy}{dx} - 6y^2\frac{dy}{dx} = 0$$

2. **Rearrange to find the gradient.**

$$\frac{dy}{dx} = \frac{2x+3y}{6y^2-3x}$$

However, for this one, you want $\frac{dx}{dy}$, which is $\frac{1}{\left(\dfrac{dy}{dx}\right)}$:

$$\frac{dx}{dy} = \frac{6y^2-3x}{2x+3y}$$

3. **Substitute in your known value for the derivative.**

You know $\frac{dx}{dy} = 0$, so $6y^2 - 3x = 0$, and $x = 2y^2$. This gives you a relationship between the coordinates where the derivative has the value you want. (Look at Figure 18-6 to see how it works.)

4. **Solve this simultaneously with the original equation.**

The points you're interested in lie on the original curve and satisfy the result from Step 3, so simultaneous equations are the way forward. Replacing x with $2y^2$, you get $4y^4 + 6y^3 - 2y^3 = 8$, or $y^4 + y^3 - 2 = 0$. There's a relatively obvious solution at $y = 1$ and, correspondingly, $x = 2$.

There is one other real root, near $(4.766, -1.544)$, but that's not in the first quadrant; the question asked for the point, not the points, so there's no need to look for it.

This is a harder question than you might expect in Core 4, but it's technically fair game. You would be led through it instead of being expected to solve it from scratch.

Figure 18-6:
The implicitly defined curve $x^2 + 3xy - 2y^3 = 8$ (solid line) and the curve $x = 2y^2$ (dashed line), on which the points with a vertical tangent have to lie.

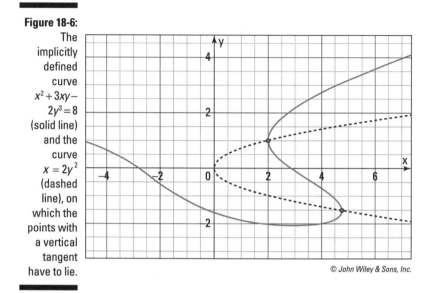

© John Wiley & Sons, Inc.

Volumes of Revolution

You use integration all the time to work out the area of a shape. In this section, you take that idea into the next dimension and use integration to work out the volume of a nice, rotationally symmetric shape.

To do this, you need to understand what's going on when you integrate to find something like $\int y\,dx$. As you can see in Figure 18-7a, that expression represents a number of extremely narrow rectangles, each of which is y tall and dx – an infinitesimally small distance – wide. The squiggly $\int$ means 'add up all these rectangle areas'.

If, on the other hand, you want to find the volume of the cone in Figure 18-7b, you could slice it up vertically into tiny *cylinders* or disks. Each of them has a radius of y and (speaking very loosely) a height of dx, meaning they each have a volume of $\pi y^2\,dx$. When you're looking for a volume of revolution, you need to integrate the square of the function you're looking at and multiply it by π.

The volume of a solid of revolution formed by rotating the curve $y = f(x)$, between the limits of a and b, by 2π radians around the x-axis is $\int_a^b \pi \left[f(x) \right]^2 dx$.

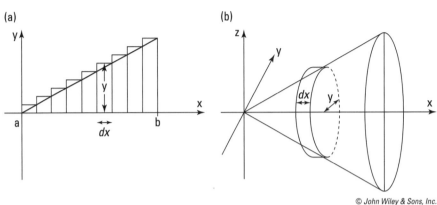

Figure 18-7:
(a) Splitting a 2D integral into tiny rectangles; (b) splitting a volume of revolution into tiny disks.

© John Wiley & Sons, Inc.

If you struggle to imagine this, that's fine – thinking in 3D is tricky – but I find it helps to think about holding a ruler up against the slanted side of an ice-cream cone and rotating the cone around its axis. The ruler stays jammed up against the side of the cone (probably getting a little sticky and raspberry-flavoured in the process).

The cone in this example is sideways-on rather than upright, and it's solid rather than empty, but the same idea applies. You're going to split it up into vertical slices, each of which is a thin cylinder. The volume of these is easy to work out (especially in comparison to, say, horizontal slices, which are probably hyperbolas with less-than-lovely equations).

Suppose the equation of a line is $y = 2x$ and you're looking at x-values from 0 to 5. Here's how to work out the volume when $y = 2x$ is spun around the x-axis, using the formula $V = \int_a^b \pi y^2 \, dx$.

1. **Substitute for y in the integral and find your limits.**

   Because $y^2 = 4x^2$ and the limits are from 0 to 5, the integral becomes $\int_0^5 4\pi x^2 \, dx$.

2. **Integrate this with respect to x.**

   $$\left[\frac{4}{3} \pi x^3 \right]_0^5$$

3. **Put in your limits and subtract.**

   The upper limit for $x = 5$ gives you $\frac{500}{3} \pi$, and the lower limit gives you 0. The difference is $\frac{500}{3} \pi$.

Because π is a constant, you can bring it out of the integral at the start and work out $\pi\int_0^5 4x^2\,dx$ instead, if you prefer. Personally, I'd leave the π in the integral so that I wouldn't forget to put it back in later!

With a cone, you can check that your answer fits the formula: the cone's base radius is 10 and its height is 5, making its volume $\frac{1}{3}\pi r^2 h = \frac{1}{3}\pi \times 100 \times 5 = \frac{500}{3}\pi$. Phew!

This example can be generalised in several ways. Changing the limits of integration on this example gives you a *frustum*, a kind of truncated cone, as you'd expect. If your line were another curve, as in the following example, you'd have something more interesting to integrate – but you'd follow the same steps.

I like to remember the simple form of the revolution integral by imagining a revolting mob walking down the street, chanting at the top of their voices, '2-4-6-8, what are we going to integrate? Square the *y*! Times by π!'

On some exam boards, you also need to know what happens if you rotate a curve around the *y*-axis rather than the *x*-axis. In that case, your disks would have a radius of *x* and a height of *dy*, making the integral $\int \pi x^2\,dy$. You would need to express *x* as a function of *y* before you tried to integrate it.

Limitless fun

It's all well and good if the examiners tell you what limits to use. But what if they ask you to rotate the region below $y = 4 + 3x - x^2$ that lies above the *x*-axis? Figure 18-8 shows this.

This problem is simple enough if you think about it for a moment: you need to find where the curve crosses the axis. That is, of course, a Core 1 problem; you can factorise the quadratic as $y = (4 - x)(1 + x)$ and figure out that the limits have to be −1 and 4. For the sake of good form, I'd better do the volume calculation, too:

$$\int_{-1}^4 \pi\left(4 + 3x - x^2\right)^2 dx = \pi\int_{-1}^4 16 + 24x + x^2 - 6x^3 + x^4\,dx$$

$$= \pi\left[16x + 12x^2 + \frac{1}{3}x^3 - \frac{3}{2}x^4 + \frac{1}{5}x^5\right]_{-1}^4$$

$$= \pi\left[\left(64 + 192 + \frac{64}{3} - 384 + \frac{1{,}024}{5}\right) - \left(-16 + 12 - \frac{1}{3} - \frac{3}{2} - \frac{1}{5}\right)\right]$$

$$= \pi\left[\frac{1{,}472}{15} + \frac{181}{30}\right] = \frac{625}{6}\pi$$

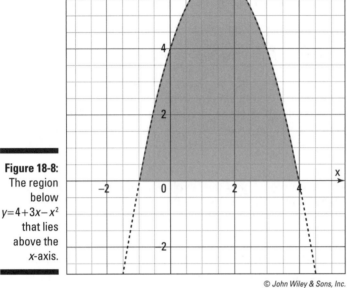

Figure 18-8:
The region below $y=4+3x-x^2$ that lies above the x-axis.

Although it's nice to do these things on paper, using a calculator here is understandable. I suppose. I do recommend working the two parts of the integral out separately on the calculator and writing them down clearly – your working should look much like what I've done here; don't just give an answer!

Combining curves

A question could reasonably combine curves for a volume of revolution in two ways. The first isn't at all tricky: it's when you have a curve defined for one set of values and a different curve for a different set that doesn't overlap. In that case, you can just work out the two volumes and add them up. For example, if you have $y = x$ for $0 \le x \le 2$ and $y = -2x^2 + 8x - 6$ for $2 < x \le 3$, and you rotate the curves around the x-axis, you find the volume for the first bit (I get $\frac{8}{3}\pi$) and for the second (which is $\frac{32}{15}\pi$) and add them up to get $\frac{24}{5}\pi$.

The second way is to look for the volume generated when the region between two curves – let's say $y_1 = f_1(x)$ and $y_2 = f_2(x)$ – is rotated around the x-axis. This is a rare-but-fair question, and you can answer it with a surprisingly simple method:

1. **Find the volume of the shape defined by one of the curves using the volume of revolution formula.**

2. **Find the volume of the shape defined by the other curve.**

3. **Find the difference between these results.**

Don't be tempted to subtract the curves and apply the formula to the result; that will give you the wrong answer, because y is squared in the integral – and $\left(y_1 - y_2 \right)^2$ (the wrong way) isn't the same thing as $y_1{}^2 - y_2{}^2$ (the right way).

Scary scalings

An unusual question (which is, all the same, fair game) involves enlarging a volume of revolution using a given scale factor. You may remember this from GCSE: if you increase all the dimensions using a scale factor of k, then the volume increases by a scale factor of k^3.

For example, if you double all the dimensions of any volume, the volume itself will be eight times as large. Similarly, multiplying each length by $\frac{1}{3}$ means the volume will be $\frac{1}{27}$ as large.

If you have a sphere with radius 2 cm, its volume is $\frac{4}{3}\pi r^3$, or $\frac{32}{3}\pi$. If you double the radius to 4 cm, the volume becomes $\frac{256}{3}\pi$, eight times as large. Finally, the volume of a cone with base radius 3 cm and height 6 cm is $\frac{1}{3}\pi r^2 h$, or 18π. If it had a base radius of 1 cm and a height of 2 cm, each a third as large, the volume would be $\frac{18}{27}\pi = \frac{2}{3}\pi$, 27 times smaller.

Of more concern to this section, if you've worked out a volume of revolution to be $\frac{24}{5}\pi$ and you're asked about the volume of a geometrically similar shape with lengths three times as large, your final volume would be $\frac{24}{5}\pi \times 27 = \frac{648}{5}\pi$.

Practising parametric volumes

A parametric curve generating a volume of revolution? I trust you will laugh at the idea, rub your hands in glee and say, 'That's . . . *trivial*.' But just in case you're not doing that for some reason, I'd better take you through the steps.

Suppose you've got the parametrically defined curve $x = \cos(t)$, $y = \sin(t)$ for $0 \le t \le \pi$, and you want to know the volume of the solid generated when you rotate it 2π radians around the x-axis. Let's go!

1. **Write down the revolution equation: $V = \int \pi\, y^2\, dx$.**

   This step might get you a mark even if you don't do anything else. It's not likely, but it might.

2. **Replace dx with something involving dt.**

   Because $\frac{dx}{dt} = -\sin(t)$, $dx = -\sin(t)\,dt$.

3. **Replace y^2 with something involving t.**

   Use $\sin^2(t)$. Your integral is now $-\pi \int \sin^3(t)\,dt$.

4. **Work out your limits in terms of t.**

   Here, the limits are the domain given to you, $0 \le t \le \pi$, so (remembering that $x = \cos(t)$) the x-values range from -1 on the left to 1 on the right. At the left-hand end, $t = \pi$, and at the right-hand end, $t = 0$. Slightly counterintuitively, the upper limit is $t = 0$ and the lower one is $t = \pi$.

5. **Do the integral!**

   $-\pi \int_{\pi}^{0} \sin^3(t)\,dt$ is best done as $\pi \int_{0}^{\pi} \sin(t)\left(1 - \cos^2(t)\right)dt$, which works out to be $\frac{4}{3}\pi$.

This answer shouldn't be too surprising. The parametric curve I set up was a semicircle with radius 1, which becomes a sphere when rotated around the axis!

Differential Equations

A little confession: throughout this chapter, I've been secretly thinking, 'Ooh! This is my *favourite topic*!' But I've been lying to myself. Differential equations are my favourite topic.

Mathematically speaking, it's a shame you don't get to go into any more depth than the separable equations you see in Core 4, although I imagine you won't thank me for saying so. If you do Further Maths or study maths at university, you're likely to spend quite a lot of time on more-involved differential equations – they're part of an incredibly rich and beautiful field. However, the main reason I like them in Core 4 is that there's a dance that goes with them.

Can you dance in the exam? Yes. Yes, you can. I've had students – even burly rugby players – do the differential equations dance in exams.

Here's how it goes: start with your hands together in front of you, as if you're clapping.

1. **Separate.**

   Pull your hands apart.

2. **Integrate.**

   Put them back together.

3. **Substitute.**

   The complicated bit: roll one hand around the other like a football manager making a substitution.

4. **Celebrate.**

   Hold them aloft like the *Y* in *YMCA*.

By a bizarre coincidence, those are *exactly* the steps you take when you're solving a differential equation. As usual with these things, the devil is in the detail.

A differential equation is any equation with a derivative in it – for example, $\frac{dy}{dx} = \frac{3}{y^2 \cos(x)}$. You may also be given a pair of values – for example, $y = 1$ when $x = 0$, or similar.

Now, nobody's told you this, because it's a great big secret: you've been solving differential equations since about Year 9. Honestly. When a question said 'the gradient of a line is 3 and it passes through the point $(4, 7)$', you were, strictly speaking, solving the differential equation $\frac{dy}{dx} = 3$, given that $y = 7$ when $x = 4$. Typically, you'd do the steps I'm about to show you, although not in so many words: you'd separate the *dy* from the *dx* and integrate: $\int dy = \int 3\,dx$, so $y = 3x + C$. That's the *general* solution – it involves a constant. You'd then substitute in your values for *x* and *y* to say $7 = 3(4) + C$, so $C = -5$ and the equation of the line is $y = 3x - 5$. That's a *particular* solution, as it finds a value for *C*.

In the following sections, I show you how to do the same thing with slightly more involved differential equations.

Generating general solutions

To *solve* a differential equation in *x* and *y* means to find the most general relationship between *x* and *y* (without any derivatives in) that satisfies the given equation and any other constraints you're given. At A level, the relationship you want usually involves getting *y* in terms of *x*.

A *general solution* to a differential equation is one that involves an arbitrary constant. Whenever you've done any integration and not found the value of C at the end, you've been finding a general solution. (If you found C, that was a particular solution, which is in the next section.)

In this section, you're going to solve $\dfrac{dy}{dx} = \dfrac{3}{y^2 \cos(x)}$ using some of the dance steps from the preceding section.

1. **Separate.**

    You want to have $\dfrac{dy}{dx}$ multiplied by something with only ys in it on the left and a function of x on the right. If you multiply by y^2, you get

    $$y^2 \frac{dy}{dx} = \frac{3}{\cos(x)}.$$

2. **Integrate (with respect to x).**

    $\int y^2 \dfrac{dy}{dx} dx = \int \dfrac{3}{\cos(x)} dx$, and the left-hand side is simply $\int y^2 \, dy$ (using the chain rule). The right-hand side can become $\int 3\sec(x) dx$, so you have $\int y^2 \, dy = \int 3\sec(x) dx$. Integrate to get $\dfrac{1}{3} y^3 = 3\ln|\sec(x) + \tan(x)| + C$.

3. **Tidy up.**

    Because you're not looking for a particular solution, there's nothing to substitute in. But you can get y on its own, just for the sake of good form:

    $$y = \sqrt[3]{9\ln|\sec(x) + \tan(x)| + c}$$

    Make sure you get the c in the right place – because you've cube-rooted after adding it, it belongs *inside* the cube root, not just tacked on at the end. It's also different to the C in the preceding step, because it's been multiplied by 3 – but that just makes it a different constant. (It's not *wrong* to write $3C$, but it looks bad.)

4. **Celebrate.**

    I don't think you need any instruction on this part!

A good question that's often asked at this point is 'How come you only need one constant, if you're integrating both sides?' The reason is that there's nothing to stop you from putting (different) constants on each side of the equation – however, you could take one of the constants away from the other and reduce it to one constant. One unknown is less trouble to deal with than two!

Picking out particular solutions

Getting a *particular solution* – usually one without any unknown constants in it – is simply a case of substituting in the values you're given, working out the missing constant(s) and putting your answer into a nice form.

In the example from the preceding section – that is, $\dfrac{dy}{dx} = \dfrac{3}{y^2 \cos(x)}$ – you know that $y = 1$ when $x = 0$. Here's what to do to get the constant:

1. **Take your general solution and replace x and y with what you know.**

   Here, $y = \sqrt[3]{9\ln\left|\sec(x)+\tan(x)\right|+c}$ becomes $1 = \sqrt[3]{9\ln\left|\sec(0)+\tan(0)\right|+c}$.

2. **Solve this for c.**

   You have $1 = \sqrt[3]{9\ln(1)+c}$, or $1 = \sqrt[3]{c}$, so $c = 1$.

3. **Replace c in the general solution.**

   Your particular solution is $y = \sqrt[3]{9\ln\left|\sec(x)+\tan(x)\right|+1}$.

This solution is about as tidy as it gets. In some cases, you'll be able to simplify your final answer further.

More-involved differential equations

Most of the Core 4 differential equations you're likely to find are of the 'here you go, solve this' variety. However, some boards (and some examiners) seem to like questions that make you work a bit harder, as if the exam were supposed to challenge your mathematical thinking rather than your memory.

In this section, I take you through some of the variations they've come up with in the past, and I show you how to deal with them.

Coming up with your own model

A favourite on some exam boards is to tell you something along the lines of 'The number of people infected with a disease is n. The rate at which the disease is spreading through the population is proportional to the cube root of the number of people already infected' and ask you to suggest a differential equation that will model this. This is really straightforward if you break it down into small chunks and translate it into maths:

> ✔ 'The rate at which the disease is spreading' is $\dfrac{dn}{dt}$. 'Rate' almost always means 'the time derivative', and here it's clearly 'the number of people infected per day' or whatever time period you choose.

✔ 'Is proportional to' means '$= k \times \cdots$', or 'some constant multiplied by'.

✔ 'The cube root of the number' means $\sqrt[3]{n}$.

Putting those together gives you $\dfrac{dn}{dt} = k\sqrt[3]{n}$, which is the answer they're after.

Interpreting models

In model questions, you're quite often asked why your value can never reach a certain level – why a temperature never drops to 15 degrees Celsius, perhaps, or a population never reaches 25,000. ('But I don't like it when they say "explain" or "interpret"!' you say. Well, I don't like it when students complain about being asked stuff they're supposed to know, so I suppose we can call it quits.)

You may be given, for example, a differential equation for a town's population x after a number of years t that looks like $\dfrac{dx}{dt} = e^{3t} - 5$. After you've solved it to get $x = \dfrac{1}{3}e^{3t} - 5t + C$, you're asked why the model is inappropriate for large values of t. The problem here is that e^{3t} gets very big very quickly and will soon dwarf any sensible population.

You have two options with these kinds of questions: the first (and usually trickiest) is to substitute the value you're given into your solution and show that there's no valid solution; the second is to look at the long-term behaviour of the solution, which I cover in Chapter 5.

Verifying solutions

Instead of solving a differential equation, you're sometimes asked to verify that a given equation really is a solution for the differential equation. This is a two-step process:

1. **Differentiate the solution to make sure you can get the differential equation you're given (perhaps with some rearrangement).**

2. **Substitute the x-value you're given into the solution and make sure you get the correct y-value out.**

Verify just means 'check this works', not 'solve this'. It's a much less involved process – and you don't get any extra marks for doing it the hard way!

For example, you may be asked to show that $y = \ln\left(x + \sqrt{x^2 + 1}\right)$ is a solution to $\dfrac{dy}{dx} = \dfrac{1}{\sqrt{x^2 + 1}}$, knowing that $x = 0$ when $y = 0$. Here's what you do:

1. **Differentiate the solution.**

You get $\dfrac{dy}{dx} = \dfrac{1 + x\left(x^2 + 1\right)^{-1/2}}{x + \sqrt{x^2 + 1}}$. Yuck.

2. Rearrange to make it nicer.

Perhaps rationalise the denominator by multiplying the top and bottom by $x - \sqrt{x^2 + 1}$ to get

$$\frac{dy}{dx} = \frac{\left(1 + \dfrac{x}{\sqrt{x^2+1}}\right)\left(x - \sqrt{x^2+1}\right)}{x^2 - \left(x^2 + 1\right)}$$

Still yuck. However, the bottom is -1, so the whole thing becomes

$$\frac{dy}{dx} = \left(1 + \frac{x}{\sqrt{x^2+1}}\right)\left(\sqrt{x^2+1} - x\right)$$

Turn the first bracket into a single fraction and multiply by the second bracket:

$$\frac{dy}{dx} = \frac{\left(\sqrt{x^2+1} + x\right)\left(\sqrt{x^2+1} - x\right)}{\sqrt{x^2+1}}$$

3. Take a deep breath and keep going. Nearly there!

The top is the difference of two squares. In the first bracket, you have the square root plus x, and in the second, you have the square root minus x, which leaves you what's under the square root, $\left(x^2 + 1\right)$, minus x^2. That's very neat: it becomes 1! That gives you $\dfrac{dy}{dx} = \dfrac{1}{\sqrt{x^2+1}}$, as required.

4. You're still not done: you need to check the solution satisfies the given condition.

The condition says that $y = \ln\left(x + \sqrt{x^2+1}\right)$ is the solution that satisfies $y = 0$ when $x = 0$. When you substitute in $x = 0$, the right-hand side is $y = \ln\left(0 + \sqrt{0^2+1}\right) = \ln(1) = 0$, as required!

Remember what I said in Chapter 3 about needing to get good at your basic algebra? This is why. Although you should be able to run through all these steps as an A level mathematician, it's probably towards the top end of what they may ask you to do – you could expect around five marks for that and probably a hint about the difference of two squares.

Finding several constants

A common variation of the find-one-constant differential equation is one that has an unknown in it to begin with, which you then need to find. Your *solution* is still the relationship between y and x, but you need to specify the constants. For example, you may be told that $\dfrac{dy}{dx} = \dfrac{k}{3y^2\sqrt{x}}$ and be given two pieces of information: when $x = 0$, $y = -1$, and when $x = \frac{1}{4}$, $y = 1$. Here's how you find the particular solution:

1. **First, go through the dance steps: separate and integrate.**

   This gives you a link between y and x in terms of your currently unknown constants: $\int 3y^2\, dy = \int kx^{-1/2}\, dx$, so $y^3 = 2k\sqrt{x} + C$.

2. **Substitute each of your pieces of information into your solution, separately, to get two equations.**

   From the first, $-1 = C$, and from the second, $1 = 2k\sqrt{\frac{1}{4}} + C$, which you can write as $1 = k + C$. (Generally, if you want to find n unknowns, you need n pieces of information, but you'd be unlucky to see n greater than 2.)

3. **Solve!**

   Here, it's nearly done: $C = -1$ and $k = 2$, so the particular solution is $y^3 = 4\sqrt{x} - 1$ or, in the form $y = f(x)$, $y = \left(4\sqrt{x} - 1\right)^{1/3}$.

4. **Celebrate!**

   What's the point in doing differential equations if you can't celebrate?

Quotient-remainder and partial fractions problems

Some examiners have a mean streak so terrible, they could almost be the baddies in a Roald Dahl book. Although no notes for *The Tale of the Differential Equation* have yet been discovered among Dahl's papers, I certainly wouldn't be inclined to rule it out. If such notes did surface, they'd doubtless contain something along the lines of the following general solution:

$$\left(4x^2 - 3x - 1\right)\frac{dy}{dx} = \left(8x^3 + 6x^2 + 13x + 8\right)$$

Makes the blood run cold, doesn't it? Here's what you do:

1. **Separate and prepare to integrate.**

   $$\int \frac{dy}{dx}\, dx = \int \frac{8x^3 + 6x^2 + 13x + 8}{4x^2 - 3x - 1}\, dx$$

2. **Split the right-hand side into partial fractions.**

   This looks like $Ax + B + \frac{C}{4x+1} + \frac{D}{x-1}$ after you factorise the bottom. Working out the constants gives you $2x + 3 - \frac{4}{4x+1} + \frac{7}{x-1}$.

3. **Integrate.**

   $y = x^2 + 3x - \ln|4x + 1| + 7\ln|x - 1| + C$, which is your general solution.

4. **Celebrate!**

   Congratulations!

In this part . . .

✔ Avoid mistakes that have scuppered students in the past.

✔ Work out where to start when you don't know where to start.

Chapter 19

Ten Classic Mistakes to Avoid

In This Chapter

▶ Not making sloppy mistakes

▶ Thinking, 'What have I missed?'

▶ Looking at the marks available

*T*here's a difference between losing marks because you don't know how to do something and losing marks because you've made a blunder – although, of course, in the exam, a dropped mark is a dropped mark.

This whole book aims to cure you of the first kind of mistake by showing you how to do just about anything that might come up. And this one, short chapter is to help cure you of the blunders. I've had more than one student get a paper back after the exam and, having missed their grade by one or two points, hold their heads in shame at some of the sloppy mistakes that had cost them. Believe me, you do *not* want to be like them.

In this chapter, I review some of the commonly committed minor mathematical sins so that – with any luck – you won't make them.

Giving Inexact Answers

'Of course it's exact! Look at how many decimal places I've written down!'

I'm afraid that's not an exact answer. No matter how many decimal places of π you write down, it's never exactly π. That's one of the things 'irrational number' means: the decimal expansion goes on forever, almost always without forming a predictable pattern.

Besides which, writing down loads of decimal places is a bit of a waste, don't you think? If π is your answer, just write π! If you've got e^3, write e^3! *That* is an exact answer. If somebody wants a decimal expansion, they can tap it into their calculator. In the meantime, e^3 or π or $\frac{\sqrt{7}}{3}$ or whatever else comes up with surds and constants in it is – for pretty much any mathematical purpose – easier to work with than the decimal your calculator gives.

More to the point, if the examiners ask for an exact answer and you give a decimal approximation, you can kiss at least one mark goodbye.

Missing Out the Constant

One of my early students had a superhero alter-ego: her classmates called her Constant Girl because, whenever they did an integration question, she'd be the one who shouted out 'plus a constant!' Obviously, it's not a patch on flying or X-ray vision, but as superpowers go, that's not a bad one to have.

Unless you're integrating between limits, you should *always* add a constant after integrating. Some questions subtly remind you of it by asking you for a curve through a given point, but if you miss the $+ C$, you can expect to lose a mark unless you're very lucky. (Some mark schemes suggest to the marker that they should condone a missing constant. These mark schemes are rare. Don't take the risk.)

Losing a Minus Sign

I must have been a horrible person to have in your maths class as a teenager. I was the kind of kid who'd routinely get scores in the mid or high 90s and be disappointed about the dropped marks. The dropped marks invariably came from sign errors.

Eventually, I became obsessive about avoiding them. Whenever I saw a bracket I needed to take away, I'd put a big warning sign beside the line in question and make sure I double-checked it at the end of the exam.

Here are some other places where it's easy to misplace a minus sign:

- When integrating negative powers or when integrating something with a negative sign in a bracket
- In binomial expansions, when multiplying several negatives together
- When integrating by parts, taking away the second integral
- Anywhere you've crammed your work into too small of a space; if you can't read it, you can't check it

Going the Wrong Way

There are a few situations where it's possible to 'go the wrong way'. For example, if you're given $f'(x)$, you may have the Pavlovian reaction of differentiating. That's what $f'(x)$ means, right?

Well, yes – but if you're given $f'(x)$, that's something that's already been differentiated. If you need $f''(x)$, then absolutely differentiate again; if you need $f(x)$, you need to integrate.

Similarly, I've seen students start from $\frac{x}{3} = 4$ and get $x = \frac{4}{3}$ (it should be $x = 12$) or go from $\sin(x) = 0.5$ to $x = \sin(0.5)$ when it ought to be $x = \sin^{-1}(0.5)$.

This is all the same problem, at root: it's applying the function itself in an attempt to get rid of it instead of applying the inverse.

Missing Trig Solutions

Tap-tap-tap-tap-tap . . . the calculator gives you an answer to the trigonometry problem. You check it. Bravo! It works. Next question, right? Not right.

The thing about trig questions is that you can almost never trust your calculator without engaging your brain. You need to ask many questions:

- ✔ Is this answer in the correct domain?
- ✔ Are there any more solutions?
- ✔ If the question asks for the smallest or largest, does your answer fit the bill?

The biggest source of dropped marks, out of those three, is missing solutions. Sometimes three or four marks are available for solutions, and if you get only one of the answers, you'll get only one of the marks.

Using the Wrong Angle Measure

Repeat after me: degrees are evil and should be banned. However, some muddle-headed questions do ask you to use them, either explicitly ('give your answer to the nearest $0.1°$') or implicitly (by giving degrees in the question).

It's very easy to forget to switch your calculator into smelly degree mode for these questions and end up with correct answers in lovely radians, which are then marked wrong because of an almost fascist insistence upon degrees. Sadly, the exam is about the only time when it's not appropriate to fight fascism.

If a question asks you for answers in degrees, give your answers in degrees and then switch your calculator back to radian mode immediately afterwards. (You can hold your nose while doing the sums, if you like.) If the question asks for answers in radians, phew! You can just do your sums in radians, as is right and proper.

Falling into a Logarithmic Booby-Trap

There are two big booby-traps when it comes to logarithms. The first is failing to make sure your solutions are valid. After you solve your question, you need to go back to the original equation and make sure your x (or whatever variable) works – remember, you're not allowed to put anything that's not positive into a logarithm.

The second booby-trap is much subtler, and it involves inequalities. I bring it up because it's a question I always used to have to fudge and only recently figured out what I was doing wrong. The problem comes when you have something like this:

$$n \ln(0.2) < \ln(0.005)$$

Naturally, you divide both sides by $\ln(0.2)$, and I would understand it if you wrote this:

$$n < \frac{\ln(0.005)}{\ln(0.2)}$$

However, it's wrong. The trouble is that $\ln(0.2)$ is negative (because $\ln(1) = 0$, and the logarithm of anything smaller than 1 is negative). Dividing by a negative number changes the direction of the inequality. That means the correct inequality is

$$n > \frac{\ln(0.005)}{\ln(0.2)}$$

Related: don't round if you're looking for the smallest integer n that satisfies the inequality – you need to use the next whole number up. Here, $n > 3.292$, and the smallest n that works is 4.

Ignoring the Rules

If I had a penny for every time someone said, 'Differentiate $x^2 \sin(x)$...
That's $2x \cos(x)$,' well, I could probably afford an extra cup of coffee or
something. I'd need one, after explaining 'no, that's not how it works' yet
again. Integrating and differentiating are easy on things added together or
taken away, but they're much harder on things multiplied or divided. That's
why you spend so much of Core 3 looking at the rules for that sort of thing.
(You can read more about the product and quotient rules in Chapter 17.)

In a similar vein, if you make mistakes like saying $(a+b)^2$ is the same thing
as $a^2 + b^2$, or that $\cos(A+B) = \cos(A) + \cos(B)$, or that $\sqrt{x^2 + y^2} = x + y$, your
examiner is going to either cry for you or laugh at you. Make sure you simplify things carefully!

 If you've fallen into the habit of getting this sort of thing wrong, try marking
problem areas with a star and making sure you check carefully before you
turn in your work. I know checking your work adds a few minutes to your
homework, but it can save you the embarrassment of seeing the red cross
and the dreaded word 'careful!' (or even 'no!') when you get your work back.

Mixing Up the Bits of a Vector Line

This isn't necessarily the most common mistake, but it's one I'm asked about
a lot in Core 4. If the vector equation of a line is $\mathbf{r} = \mathbf{a} + \lambda\mathbf{b}$, then $\mathbf{a}$ is the position vector of a reference point on the line and $\mathbf{b}$ is the direction vector. You
can tell this because as λ changes, you move by multiples of $\mathbf{b}$.

You can also draw an analogy with the traditional $y = mx + c$ version of a
straight line: the thing on its own is the reference point (c, or rather, $(0, c)$ is
a point on the line), and the thing with a variable is the gradient – the direction you're moving in.

It's also easy to mix up what's going on in two different lines. If, like me, you
didn't have the option of Greek GCSE, you may find yourself confusing the
unfamiliar λ and μ in a way you wouldn't with s and t, for example. There's
nothing to stop you from swapping the letters for ones you're more comfortable with, though!

For more info on the vector form of line, see Chapter 13.

Losing Track of the Letters

When you look at an implicitly defined curve (see Chapter 18), it's quite reasonable to find yourself a bit perplexed by the sheer number of letters involved – especially if the examiners have thrown in a constant or two for good measure. Keeping track of what you're solving for can be genuinely tricky.

The key to this sort of question is neat layout and being extra careful about what's on each side of the equation. Make sure you know the difference between $\frac{dy}{dx}$, $\frac{dy}{dt}$ and $\frac{dx}{dt}$ – and don't get any of them mixed up with x, y or t!

Chapter 20

Ten Places to Start When You Don't Know Where to Start

*N*o matter how well-prepared you are for an exam, you sometimes look at something and say, 'I have *no idea* where to start with that one.' It's happened to me, and I bet it's happened to Brian Cox as well. He got a D in his A level maths, incidentally. The difference between someone like you, aiming for a top grade, and someone like young Coxy, just scraping through, is that you have strategies for dealing with such questions.

What's that? You *don't*? Oh dear. Better fix that, eh? Luckily, there are ten of them in this chapter.

Swearing like a Trooper

This approach isn't necessarily one for the exam hall, unless you can swear like a trooper under your breath. That said, having a good, foul-mouthed rant at the question, the fool who set it, yourself for picking this idiotic subject and me for writing the stupid book is quite a good way to burn off a bit of excess energy and focus your mind.

You can explain in glorious, expletive-laden detail what exactly you find hard about the question – and what you might do with it if you weren't busy swearing at it – and eventually trick yourself into answering it.

Another possibility – again, not for the exam hall – is to get up and move around a bit. Jump up and down, put some music on and dance like nobody's watching for a few minutes, walk up and down the stairs – do something to get the old blood flowing to the brain!

If you're really stuck: take a break and do something else for 20 minutes. Your brain will work on the problem while you're otherwise engaged. It's not unusual to come back to a problem later and wonder what you found so difficult!

Sitting Up Straight and Breathing

Have you ever watched someone who's clearly uncomfortable answering a question? They're hunched over, and their breaths are shallow. That makes it quite hard for their brain to get the oxygen it needs to function.

If you find yourself hunched over, you'd do well to sit up straight and breathe deeply – get the air deep into your lungs. Deep breathing isn't just about fuelling your brain, although that's vital; it's also about trying to get yourself into a state of mind where you're thinking about the question rather than how uncomfortable and panicky you are.

If you're regularly experiencing the symptoms of panic, please go see a doctor about it. I lived with undiagnosed panic attacks for many years and caused myself a lot of unnecessary suffering.

Related: students who say things like 'I can do this. I just need to figure out what I'm missing' tend to do *way* better than students who say 'I'm rubbish at maths. I'll never get it.' Be positive about yourself – at the very least, it'll make you feel better than talking yourself down.

Making an Information Checklist

I *love* checklists. If I didn't have checklists, nothing would ever get done, and this book would probably still be in production. Making a list of what you know and what you think you need is also one of the most powerful ways that I know of to tackle a maths problem.

Here's how this approach works for a word problem:

1. **Skim through the question once to get an idea of what it's about.**

2. **Read through it again, carefully, line by line, and make notes.**

   Write down every piece of information that looks useful. Give every number a name, and make a note of every technical word you're given.

3. **Ask, 'What information do I need?'**

   For example, if you're trying to find the equation of a line, you need the gradient and a point on the line. Are you given either of those? Can you figure them out?

If nothing else, if you have to leave the question and come back to it later, you'll have all the information neatly summarised.

Putting Information Together

One of the best questions you can ask yourself when faced with an impossible question is 'How can I combine the information I've got?'

If you know you're working with a triangle and you have some sides and/or angles, is there a rule you can use to work out the rest? Even if that info doesn't immediately help, it may spark something. If you're midway through a question and something odd crops up, you may ask, 'How can I relate this to something I've already worked out?' You may even think, 'What if I made those two things equal? Would that make any sense?' Often it doesn't, and you need to try something else, just as if you were putting a jigsaw puzzle together.

Drawing a Big Diagram

I once had a teacher who insisted that if a drawing took up less than half a page, it didn't count as a diagram. I think that's slightly excessive, but at the same time, paper is there to be used, and the bigger your diagram, the easier it is to see your labels.

Seeing what a problem looks like means you can think about it in slightly different ways. For example, in a geometry problem, you may be able to see that one of the possible answers you came up with doesn't make sense, or (if you're lucky) you may get an estimate of what your answer ought to be.

If you're working with a triangle, drawing a more-or-less-accurate scale diagram will often give you an idea of what kind of number to expect. By looking at the picture, you can get a rough sense of what angle or length you're after.

Your diagram doesn't need to be a Rembrandt; it's not going in the Tate. Just draw big, draw clear, and draw fast. For more on sketching graphs, see Chapters 4 and 10.

Rereading the Question

They try to bully you, those exam boards do – intimidate you with big blocks of text, hoping you'll cower rather than read through all the details. Well, don't let them. You're stronger than that. Go back through the question, carefully, as if you were a detective on the lookout for clues. Ignore anything irrelevant, but pay close attention to anything that looks helpful.

If you can get started, great! Go for it. If not, go through the question again and see whether you missed anything. Always be thinking, 'What information would be useful for me to try to solve this puzzle? What's missing?'

Look back earlier in the question, too, to see whether there's anything by way of a hint. Frequently a 'show that...' in part (a) will be recycled for use again in part (b) or (c)!

The more words there are in a question, the more likely it is they're trying to help you through it. It's just a case of teasing out the hints they've given you!

Starting at the End

Skipping to the end and working backwards is a particularly good thing to do with a proof question when you've analysed yourself into a hole. It's also handy for those silly Core 3 numerical-methods questions where you have to rearrange a nice equation into an ugly one for the sake of making an iteration converge (see Chapter 9 for details).

After you've taken one side of a proof as far as you can, work on the other side, too. You more or less make a fresh start at the end of the question and work backwards until you see a way to bridge the gap between the two sides.

It's a good idea to rewrite your work in the correct order, though, as if you'd gone straight through from start to finish. Even if you're using sneaky work-arounds, you probably don't want to *look* as though you are!

A ninja trick that normally works well for the numerical-methods questions is to write where you're trying to get to at the bottom of the page and work upwards until you get to where you're meant to start. That's often less complicated that going forwards, surprisingly!

Starting with the Ugliest Thing

You may disagree with me, but I believe maths is about making things beautiful. (If you disagree with me, you'd be wrong, by the way.) That's not just a statement of aesthetics; it's also a good tip for how to approach many A level maths problems: find the ugliest thing and make it less ugly.

Integrating with something awful in a bracket? Try calling that *u* (for *ugly*). Got a square root that doesn't simplify away? Try squaring both sides of the equation. (Be careful not to introduce spurious solutions! It's always a good idea to make sure your solutions actually answer the question you were asked.)

That fraction with fractions on the top and on the bottom? Try multiplying both top and bottom by something to make them nicer. And then probably multiply both sides of the equation by the bottom of the fraction so everything you're working with is on the top line – while there are no guarantees, it's worth trying and seeing where you end up!

That monster that's come out of the quotient rule with e^xs all over the place? Try factorising those out.

You may need a bit of study to figure out what constitutes ugliness, but after you get a sense of mathematical neatness, an awful lot of nasty things will drop into place.

Ignoring the Hard Bit

There's a category of questions known – among A level aficionados – as *bloodbath questions*. They're usually towards the end of the paper and have a load of marks associated with them. Sometimes it's 15 marks scattered across five or six parts; sometimes it's just a 9 in brackets. And they're the kinds of questions students end up making memes of to complain about.

Even though they're perfectly fair questions, they may involve ideas you haven't seen in a while. For example, a famous Core 3 bloodbath required students to know without prompting that $\sin(90° - x) \equiv \cos(x)$. Or the question may be presented in an unusual way. Another bloodbath question used a silly setup involving a photographer and a load of background that was completely unnecessary for what turned out to be a straightforward trigonometry problem – if you read that far.

Often with bloodbath questions, you can pick up marks even if you don't spot everything straight away. It's tempting to just give up, but that's a good recipe for throwing away a load of marks – and losing 15 marks could drop you as much as two full grades in a paper.

A better plan is to pick bits you *can* do – probably using information from earlier parts of the question. Even if you can only go so far with your answer, you can still get partial credit.

Giving up is rarely the correct answer.

Asking, 'What Would Colin Ask?'

Imagine (you can do the wibbly vision thing if you want to): you're in class with me or another teacher who knows his or her stuff. Imagine that teacher is watching you try to solve the question and seeing you struggle.

What would this person ask you? What hints might you expect to get? What would make this teacher roll his or her eyes and sigh?

It's a weird but powerful idea: thinking about what an expert would do, oddly, can help turn you into an expert. Looking at things from someone else's point of view can be the push you need to get started!

Index

About the Author

Colin Beveridge is the author of four other *For Dummies* titles:

- *Basic Maths For Dummies*
- *Basic Maths Practice Problems For Dummies*
- *Numeracy Tests For Dummies*
- *Teachers' Skills Tests For Dummies* (with Andrew Green)

and the forthcoming popular maths book *Cracking Mathematics.*

Once a researcher for NASA's *Living With a Star* program, Colin left academia in 2008 to become a tutor, speaker and writer, and he is still waiting for the job police to catch up with him and make him do something less fun.

He lives in Weymouth, Dorset, with his family, and he spends his days looking after young Bill and Fred or training to run long distances marginally less slowly. He also speaks regularly at his local Toastmasters club and rarely misses a MathsJam.

Dedication

This one's for my dad, Ken, who told me I should always do my best work.

Author's Acknowledgments

A huge thank you to Nicky Russ, who dealt with childcare while I dealt with deadlines, and to everyone else who stepped up to help out at critical moments – especially Laura Russ and Linda Hendren.

I'd never have been in a position to write this book without bouncing ideas off of hundreds of students over a decade or so: special thanks to Becky Bailey, Katlyne and Coralie Scannell, Hugo Rowland and Dominika Vasilkova.

Lastly, the crew at Dummies Towers have done their usual stellar job. I'm particularly grateful to Chrissy Guthrie, Danielle Voirol and E. A. Williams for catching my mistakes, telling me off and putting them right.

Publisher's Acknowledgments

Executive Commissioning Editor: Annie Knight

Editorial Project Manager and Development Editor: Christina Guthrie

Copy Editor: Danielle Voirol

Technical Editor: Elizabeth A. Williams

Production Editor: Antony Sami

Cover Image: ©Getty Images/blackred

Take Dummies with you everywhere you go!

Whether you're excited about e-books, want more from the web, must have your mobile apps, or swept up in social media, Dummies makes everything easier.

FOR DUMMIES®

A Wiley Brand

BUSINESS

978-1-118-73077-5

978-1-118-44349-1

978-1-119-97527-4

MUSIC

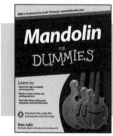

978-1-119-94276-4

978-0-470-97799-6

978-0-470-49644-2

DIGITAL PHOTOGRAPHY

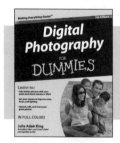

978-1-118-09203-3

978-0-470-76878-5

978-1-118-00472-2

Algebra I For Dummies
978-0-470-55964-2

Anatomy & Physiology For Dummies, 2nd Edition
978-0-470-92326-9

Asperger's Syndrome For Dummies
978-0-470-66087-4

Basic Maths For Dummies
978-1-119-97452-9

Body Language For Dummies, 2nd Edition
978-1-119-95351-7

Bookkeeping For Dummies, 3rd Edition
978-1-118-34689-1

British Sign Language For Dummies
978-0-470-69477-0

Cricket for Dummies, 2nd Edition
978-1-118-48032-8

Currency Trading For Dummies, 2nd Edition
978-1-118-01851-4

Cycling For Dummies
978-1-118-36435-2

Diabetes For Dummies, 3rd Edition
978-0-470-97711-8

eBay For Dummies, 3rd Edition
978-1-119-94122-4

Electronics For Dummies All-in-One For Dummies
978-1-118-58973-1

English Grammar For Dummies
978-0-470-05752-0

French For Dummies, 2nd Edition
978-1-118-00464-7

Guitar For Dummies, 3rd Edition
978-1-118-11554-1

IBS For Dummies
978-0-470-51737-6

Keeping Chickens For Dummies
978-1-119-99417-6

Knitting For Dummies, 3rd Edition
978-1-118-66151-2

FOR DUMMIES®

A Wiley Brand

SELF-HELP

978-0-470-66541-1

978-1-119-99264-6

978-0-470-66086-7

LANGUAGES

978-0-470-68815-1

978-1-119-97959-3

978-0-470-69477-0

HISTORY

978-0-470-68792-5

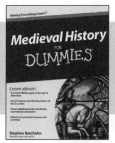

978-0-470-74783-4

978-0-470-97819-1

Laptops For Dummies 5th Edition
978-1-118-11533-6

**Management For Dummies,
2nd Edition**
978-0-470-97769-9

Nutrition For Dummies, 2nd Edition
978-0-470-97276-2

Office 2013 For Dummies
978-1-118-49715-9

Organic Gardening For Dummies
978-1-119-97706-3

Origami Kit For Dummies
978-0-470-75857-1

Overcoming Depression For Dummies
978-0-470-69430-5

Physics I For Dummies
978-0-470-90324-7

Project Management For Dummies
978-0-470-71119-4

Psychology Statistics For Dummies
978-1-119-95287-9

**Renting Out Your Property For Dummies,
3rd Edition**
978-1-119-97640-0

Rugby Union For Dummies, 3rd Edition
978-1-119-99092-5

Stargazing For Dummies
978-1-118-41156-8

**Teaching English as a Foreign Language
For Dummies**
978-0-470-74576-2

Time Management For Dummies
978-0-470-77765-7

Training Your Brain For Dummies
978-0-470-97449-0

Voice and Speaking Skills For Dummies
978-1-119-94512-3

Wedding Planning For Dummies
978-1-118-69951-5

WordPress For Dummies, 5th Edition
978-1-118-38318-6

Think you can't learn it in a day? Think again!

The *In a Day* e-book series from *For Dummies* gives you quick and easy access to learn a new skill, brush up on a hobby, or enhance your personal or professional life — all in a day. Easy!